Comparative Entrepreneurship Education

Xiaozhou Xu
Editor

Comparative Entrepreneurship Education

Springer

Editor
Xiaozhou Xu
College of Education
Zhejiang University
Hangzhou, China

ISBN 978-981-99-1834-8 ISBN 978-981-99-1835-5 (eBook)
https://doi.org/10.1007/978-981-99-1835-5

This Springer imprint is published by the registered company Springer Nature Singapore Pte Ltd.
The registered company address is: 152 Beach Road, #21-01/04 Gateway East, Singapore 189721, Singapore

Contents

Contributors

Aisha Siddiqa Husain is a PhD student in Higher Education at the Ontario Institute for Studies in Education of the University of Toronto, Canada. Her research focuses on entrepreneurship education.

Arthur K. Ellis is the Director of the Center for Global Curriculum Studies at Seattle Pacific University, USA. Prior to this, he was a Full Professor of Education at the University of Minnesota. An author of over 20 books and numerous journal articles, he was a consultant for projects by the National Science Foundation and the American Association for the Advancement of Science. In addition to his doctorate from the University of Oregon, he holds two honorary doctorates.

Christine Volkmann is Chair of Entrepreneurship and Economic Development, and UNESCO Chair of Entrepreneurship and Intercultural Management at the Schumpeter School of Business and Economics, University of Wuppertal, Germany. She is also the executive spokeswoman of the interdisciplinary Jackstädt Research Center for Entrepreneurship and Innovation and co-founder of a university-wide, regional innovation lab.

Creso Sá is a Distinguished Professor of Science Policy, Higher Education, and Innovation at the University of Toronto, Canada. He is also the Director of the Center for the Study of Canadian and International Higher Education, and Associate Chair of the Leadership, Higher, and Adult Education Department. His research interests include science and technology policy, the organization of science, and the evolving role of universities in supporting innovation.

Jarna Heinonen is a Professor of Entrepreneurship at the Department of Management and Entrepreneurship, University of Turku, Finland. Her research interests include entrepreneurship, particularly entrepreneurship education, corporate entrepreneurship, organizational renewal, family businesses, and related practices and policies. Widely published in scholarly journals, she is highly experienced in building bridges between academia and business life—a "pracademic" collaborating with businesses as a researcher, teacher, and adviser.

Jeremy Delamarter is an Associate Professor of Education at St. Martins University, USA.

Leanna Aker is a Professor of Education and Director of Field Placement at the City University of Seattle, USA.

Marc Grünhagen is a lecturer and researcher in entrepreneurship at the Schumpeter School of Business and Economics, University of Wuppertal, Germany. Prior to joining the entrepreneurship team at the University of Wuppertal, he worked as a consultant for technology ventures and as an investment manager for a small venture capital fund.

Robin Bell is a Professor of Entrepreneurship and the DBA Programme Director at the Worcester Business School. He led the development of the Worcester Business School's first online degree programme, and fostered new relationships and grew existing partnerships to develop collaborative programmes, student progression, research, and mobility agreements, as the Director of International Partnerships. His innovative curriculum development and teaching approaches to the field of entrepreneurship and enterprise have been recognized by multiple awards, and he has been the recipient of research and teaching funding from a wide range of institutions. Robin was previously a Visiting Associate Professor at Beijing Foreign Studies University and has advised on development of numerous entrepreneurship education programmes and ecosystems across China, Europe, and Africa.

Sanna Ilonen is a lecturer in entrepreneurship at the Department of Management and Entrepreneurship, University of Turku, Finland. She is an active member of the scientific community in the field of entrepreneurship education and serves as a deputy board member for the Scientific Association for Entrepreneurship Education.

Slavica Singer is a Professor Emeritus of Entrepreneurship at the J.J. Strossmayer University in Osijek, Croatia. She is a pioneer of research-based entrepreneurship education in Croatia and was a board member of the Global Entrepreneurship Research Association (2010–16). She serves as UNESCO Chair in Entrepreneurship Education and was awarded the European Entrepreneurship Education Award in 2021.

TaeUk Ahn is a Research Assistant Professor of Entrepreneurship at KAIST (Korea Advanced Institute of Science and Technology) in South KOREA. He is a PhD in Entrepreneurship and has a lot of experience in startups and venture company. He served as a professor of Entrepreneurship at Kunsan National University and the head of the local Government's Entrepreneurship Center. He was awarded the Republic of Korea Venture Startup Promotion Merit in the Ministry of SMEs and Startups in 2020.

Weihui Mei is a Professor of Comparative Education at the College of Education, Zhejiang University, China. Her research interests include entrepreneurship education, comparative higher education, university and regional development. Her research has been funded by grants from bodies such as UNESCO Bangkok Office and the National Social Science Fund of China. She most recently authored Innovation of Organizational Models and Operational Mechanism of Entrepreneurship Education (Zhejiang University Press, 2020).

Xiaozhou Xu is the President of Communication University of Zhejiang, China; Cheung Kong Scholars Chair Professor of China's Ministry of Education; and Qiushi Chair Professor of Zhejiang University. He also serves as UNESCO Chair of Entrepreneurship Education and President of the China Chapter of UNESCO Entrepreneurship Education Network.

List of Figures

List of Tables

Chapter 1
Introduction

Xiaozhou Xu and Weihui Mei

During the first World Conference on Higher Education in 1998, the United Nations Educational, Scientific and Cultural Organization (UNESCO) proposed that "Developing entrepreneurial skills and initiative should become major concerns of higher education, in order to facilitate employability of graduates who will increasingly be called upon to be not only job seekers but also and above all to become job creators" (UNESCO, 1998). Since then, along with the sustained promotion of international organizations and national policies, entrepreneurship education has been widely established, not only in developed countries but also in emerging countries, to promote economic prosperity, social mobility, and sustainable development (Mei & Symaco, 2022; Sá & Kretz, 2015). This book discusses the necessity of entrepreneurship education against the backdrop of the pandemic and the Fourth Industrial Revolution; compares the similarities and differences of entrepreneurship education in selected countries; and outlines the future of entrepreneurship education from a global perspective.

1.1 Background of Entrepreneurship Education

The past two decades have seen tremendous technological progress, knowledge advancement, higher education expansion, and great social challenges, all of which required joint efforts from different countries.

X. Xu (✉) · W. Mei
College of Education, Zhejiang University, Hangzhou, China
e-mail: xxiaozhou@163.com

X. Xu, *Comparative Entrepreneurship Education*,
https://doi.org/10.1007/978-981-99-1835-5_1

First, the emergence of Fourth Industrial Revolution calls for innovative and entrepreneurial talents.

With the convergence of digital, physical, and biological technologies, a global transformation is taking place—and it is historic in terms of its velocity, breadth, and depth, and systems impact (Schwab, 2017). The World Economic Forum (WEF) estimates that "by 2025, 85 million jobs may be displaced by a shift in the division of labour between humans and machines, while 97 million new roles may emerge that are more adapted to the new division of labour between humans, machines and algorithms" (WEF, 2020). The year 2020 is especially challenging because "automation, in tandem with the COVID-19 pandemic, is creating a 'double-disruption' scenario for university graduates and other workers" (WEF, 2020).

However, technological progress can also bring new opportunities for university graduates. It can lower the barriers to startup activities and reduce obstacles from invention to the market (Sá & Kretz, 2015, p. 6). In addition, the development of digital technology is conducive to the elimination of geographical barriers, which can provide a global market for entrepreneurs (Zhao, 2020). In the era of digital intelligence, new business models are emerging, existing business models are subverted, and production, consumption, transportation, and delivery systems are reshaped (Schwab, 2017). The digitalization, networking, and intellectualization of this new industrial revolution will give birth to new business forms, models of employment, and entrepreneurial activities (Zhao & Xu, 2020).

Technological progress can also blur the boundary between formal and informal work (World Bank, 2019). It is estimated that there are around 84 million freelancers globally, and more than two-thirds of the 57.3 million freelancers in the United States also hold a traditional job, using freelancing to supplement their income (World Bank, 2019). Therefore, in the era of the digital economy, anyone with real talent do not need to adhere to the requirements of the traditional labor market. Rather, they can rely on the digital platform to balance their formal and informal work, so as to enhance personal and social value (Mei, 2020a).

All these changes have shaped a world of volatility, uncertainty, complexity, and ambiguity, and have brought both opportunities and challenges for university graduates. Schools and higher education institutions (HEIs) should reconsider the knowledge and skills needed and "prepare [students] for jobs that have not yet been created, for technologies that have not yet been invented, to solve problems that have not yet been anticipated" (OECD, 2018). As pointed out in the World Bank's *The Changing Nature of Work* (2019), "Technology is reshaping the skills needed for work……The demand for advanced cognitive skills, socio-behavioral skills, and skill combinations associated with greater adaptability is rising". Therefore, countries should reinvest in education, especially cultivating students' key competencies that are needed to better cooperate with machines or that machines cannot compete with such as critical thinking, creativity, innovation and entrepreneurship, interpersonal communication ability (e.g., leadership and teamwork), active learning ability, and social emotional ability (e.g., empathy) (Mei, 2020b).

Second, the United Nations' 2030 Agenda for Sustainable Development has put entrepreneurship in a strategic position.

The Sustainable Development Goal (SDG) 4.4 emphasizes that by 2030, we should "substantially increase the number of youth and adults who have relevant skills, including technical and vocational skills, for employment, decent jobs and entrepreneurship" (United Nations, 2015). Education systems should transform to make entrepreneurship education more inclusve (OECD, 2019).

Innovation and entrepreneuship have also been regarded as an effective way to alleviate global poverty and promote economic prosperity. SDG 8.3 requires all countries to "promote development-oriented policies that support productive activities, decent job creation, entrepreneurship, creativity and innovation, and encourage the formalization and growth of micro-, small- and medium-sized enterprises, including through access to financial services" (United Nations, 2015). For example, China is making full use of technology and entrepreneurship to promote underdeveloped rural areas. HEIs should effectively provide social entrepreneurship education (EE) to cultivate students with both entrepreneurial competencies and social responsibilities, so as to tackle challenges such as inequality, poverty, pollution, and public health.

Third, the study of EE has developed very quickly in the past two decades globally, but theoretical and comparative studies are quite limited.

Using bibliometric software CiteSpace, we performed a visualization analysis of 692 articles from the Web of Science Core Collection published in 2000–2019. The results show that the number of international EE literature is on the rise in the past 20 years. Based on keyword cluster analysis, we identified eight themes: creativity, entrepreneurial intention, learning technologies, commercialization, soft skills, regional context, identity, and social entrepreneurship (see Fig. 1.1). We can see that contemporary studies lack the analysis of various contexts, diversified models, and the key challenges of EE in different countries, which are of great importance to further the development of EE. Volkmann and Audretsch (2017) compared the entrepreneurial activities of 20 HEIs in different European countries, which sets a good example for further study. This book extends the countries being analysed to America and Asia, with the former having the longest history of EE, and the latter with the biggest higher education system in the world. It attempts to systematically analyze the approaches, achievements, and challenges of EE in different countries, to share best practices, and to provide recommendations for the future development of EE. The comparative perspective can aid policymakers, researchers, and students in examining EE in more comprehensive and far-reaching ways.

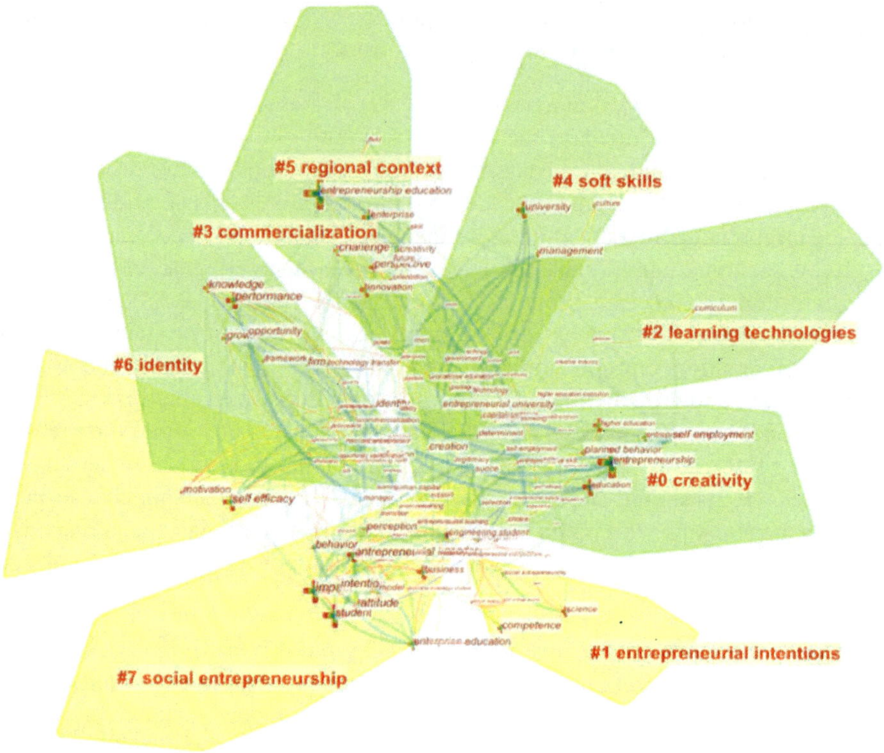

Fig. 1.1 Research themes of entrepreneurship education in 2000–2019

1.2 Selected Countries

This book selects Canada, China, Croatia, Finland, Germany, South Korea, the United Kingdom, and the United States as case countries. They were selected based on the following criteria:

- **Broad Geographic Coverage**: The countries cover America, Europe, and Asia, including the United States and Canada (North America), the United Kingdom (Northwest Europe), Finland (North Europe), Germany (Central Europe), Croatia (Southeastern Europe), China and South Korea (East Asia), etc.
- **High Impact of EE**: The EE in these countries has a long history, many students, and particularly significant impact on the economic and social development of each country.
- **Uniqueness and Innovation:** These countries have taken unique approaches to make EE more innovative and inclusive.
- **Possibility of Transferability:** The approaches are quite transparent and can be transferred to other countries.

1.3 Contributions of This Book

1.3.1 The Concept of Entrepreneurship Education

Though different organizations and individuals have tried to make a clear definition of EE, no consensual definition exists (Attali & Yemini, 2017). Generally speaking, there are four approaches to understanding entrepreneurship and EE (Mei, 2020a). The first approach recognizes entrepreneurship as a personal trait and believes that entrepreneurs and non-entrepreneurs have significant differences in personal traits (Brockhaus & Horwitz, 1986). The second approach believes that entrepreneurship is a process and the content of EE should be organized around the process of entrepreneurship. Neck et al. (2014) analyzed 45 EE texts in the United States and found that 80% of the texts emphasized the process of entrepreneurship. The third approach believes that entrepreneurship pertains to an individual's ability to turn ideas into action—a key competence—and young people should be encouraged to be more creative and confident in whatever they undertake (European Commission, 2008). The fourth approach proposes to teach entrepreneurship as a method, including starting businesses as coursework, serious games and simulations, design-based thinking, and reflective practice (Neck & Greene, 2011). This book views EE from a broad perspective and recognizes entrepreneurship as a key competence for all students; therefore, it should be embedded into the process of talent cultivation. HEIs should on the one hand adopt an approach of experiential learning to better implement entrepreneurship education, and on the other hand construct an ecosystem to support diverse entrepreneurship activities.

1.3.2 Structure of the Book

There are ten chapters of this book. The first chapter is Introduction, analyzing the background of entrepreneurship education, the reason of choosing the case countries and the contributions of this book.

Following the Introduction, there are eight chapters. The analysis is based on country-by-country descriptions of the context, polices, practices, and characteristics of EE within their national settings. Context matters. The famous comparative education pioneer Sir Michael Sadler stated in his Oxford Speech in 1900: "In studying foreign systems of education, we should not forget that the things outside the schools matter even more than the things inside the schools, and govern and interpret the things inside. We cannot wander at pleasure among the educational systems of the world, like a child strolling through a garden, pick off a flower from one bush and some leaves from another, and then expect that if we stick what we have gathered into the soil at home, we shall have a living plant. A national system of education is a living thing" (Sadler, 1900). This book aims to provide a context analysis that can lead to greater insight into why and how EE has become an important government agenda and an institutional priority in different country settings. Following the context, each chapter will analyze governmental policies and the guidance of entrepreneurship education in recent years. Each chapter will then explain how HEIs implement EE in terms of curriculum, co-curriculum, organizational structure, and so on. Special attention will also be paid to the supporting system in each country, such as how stakeholders support entrepreneurship activities, where resources come from, the role of entrepreneurship platforms, etc. Finally, the characteristics and tendencies of EE will also be analyzed. Though each chapter may use different logic or sequence of analysis, it will cover the above-mentioned content.

Based on the comparison of case countries, Chap. 10 puts forwards the common successful experience and the differentiation of EE. It is also argued that systematically support the development of EE, reforming organizational structures and supporting institutions, making EE an academic area of study as well as strengthening the global EE exchange and cooperation will be the future tendencies of global EE.

References

Attali, M. O., & Yemini, M. (2017). Initiating consensus: Stakeholders define entrepreneurship in education. *Educational Review, 69*(2), 140–157.

Brockhaus, R., & Horwitz, P. (1986). The psychology of the entrepreneur. In D. Sexton & R. Smilor (Eds.), *The art and science of entrepreneurship* (pp. 25–48). Ballinger.

Brush, C. G. (2014). Exploring the concept of an entrepreneurship education ecosystem. In S. Hoskinson & D. F. Kuratko (Eds.), *Innovative pathways for university entrepreneurship in the 21st century* (pp. 25–39). Emerald Group Publishing Limited.

Christopherson, R. W. (1997). *Geosystems: An introduction to physical geography.* Prentice Hall.

European Commission. (2008). *Entrepreneurship in higher education, especially within non-business studies.* Author. https://ec.europa.eu/docsroom/documents/8969/attachments/1/translations/en/renditions/pdf

Fetters, M. L., Greene, P. G., Rice, M. P., & Butler, J. S. (2010). *The development of university-based entrepreneurship ecosystems: Global practices.* Edward Elgar Publishing.

Isenberg, D. J. (2010). How to start an entrepreneurial revolution. *Harvard Business Review, 88*(6), 40–50.

Kingdon, J. W. (2014). *Agendas, alternatives, and public policies* (2nd ed.). Pearson.

McKinsey & Company. (2011, October). *The power of many: Realizing the socioeconomic potential of entrepreneurs in the 21st century economy.* http://www.iblfrussia.org/files/The_Power_of_Many-_McKinsey_Report-_20110310.pdf

Mei, W. H. (2020a). *Innovation of organizational models and operational mechanism of entrepreneurship education in higher educational institutions.* Zhejiang University Press [In Chinese].

Mei, W. H. (2020b, December 22). Facing digital employment and entrepreneurship: What competencies should college students strengthen? *Guangming Daily*, p. 13 [In Chinese].

Mei, W. H., & Symaco, L. (2022). University-wide entrepreneurship education in China's higher education institutions: Issues and challenges. *Studies in Higher Education, 47*(1), 177–193.

Neck, H. M., & Greene, P. (2011). Entrepreneurship education: Known worlds and new frontiers. *Journal of Small Business Management, 49*(1), 55–70.

Neck, H. M., Greene, P. G., & Brush, C. G. (2014). *Teaching entrepreneurship: A practice-based approach.* Edward Elgar Publishing.

Neck, H. M., Meyer, G. D., Cohen, B., & Corbett, A. C. (2004). An entrepreneurial system view of new venture creation. *Journal of Small Business Management, 42*(2), 190–208.

Organization for Economic Cooperation and Development, & European Union. (2019, December). *The missing entrepreneurs 2019: Policies for inclusive entrepreneurship.* OECD Publishing. https://doi.org/10.1787/3ed84801-en

Organization for Economic Cooperation and Development. (2018, April). *The future of education and skills: Education 2030.* OECD Publishing. https://www.oecd.org/education/2030/E2030%20Position%20Paper%20(05.04.2018).pdf

Regele, M. D., & Neck, H. M. (2012). The entrepreneurship education sub-ecosystem in the United States: Opportunities to increase entrepreneurial activity. *Journal of Business & Entrepreneurship, 23*(2), 25–47.

Sá, C. M., & Kretz, A. J. (2015). *The entrepreneurship movement and the university.* Palgrave Macmillan.

Sadler, M. (1900, October). How far can we learn anything of practical value from the study of foreign systems abroad? In *Address at Guildford Education Conference.*

Schwab, K. (2017). *The fourth industrial revolution.* Crown Business.

UNESCO. (1998). *World declaration on higher education for the twenty-first century: Vision and action.* Paper adopted by the World Conference on Higher Education in the Twenty-First Century: Vision and Action, Paris, France. Retrieved from https://unesdoc.unesco.org/ark:/48223/pf0000141952

United Nations. (2015). *Transforming our world: The 2030 agenda for sustainable development.* Retrieved from https://sustainabledevelopment.un.org/content/documents/21252030%20Agenda%20for%20Sustainable%20Development%20web.pdf.

Volkmann, C. K., & Audretsch, D. B. (2017). *Entrepreneurship education at universities: Learning from twenty European cases.* Springer.

Wilson, K. E., Vyakarnam, S., Volkmann, C., Mariotti, S., & Rabuzzi, D. (2009, April). *Educating the next wave of entrepreneurs: Unlocking entrepreneurial capabilities to meet the global challenges of the 21st century.* World Economic Forum. https://ssrn.com/abstract=1396704 or https://doi.org/10.2139/ssrn.1396704

World Bank. (2019). *The changing nature of work.* The World Bank Group. https://documents1.worldbank.org/curated/en/816281518818814423/2019-WDR-Report.pdf

World Economic Forum. (2020, October). *The future of jobs report 2020.* World Economic Forum. Retrieved from https://www3.weforum.org/docs/WEF_Future_of_Jobs_2020.pdf

Xu, X. Z. (2019). Strategies for constructing ecosystem of innovation and entrepreneurship education in colleges and universities in the period of transformation and upgrading. *Research in Educational Development, 39*(Z1), 102–108. [In Chinese]

Zhao, C. W., & Xu, Z. Y. (2020). *Industrial upgrading in the era of new industrial revolution in China.* Peking University Press [In Chinese].

Zhao, Y. (2020). Education in the age of smart machines: Directions and strategies. *Educational Research, 41*(3), 26–35 [In Chinese].

Chapter 2
Entrepreneurship Education in Canada

Creso Sá and Aisha Husain

Entrepreneurship education (EE) has become ubiquitous in Canadian higher education over the last two decades. This has happened in the way that change typically occurs in higher education in this country. No one planned or anticipated this expansion, nor was any national vision or direction considered. Each provincial government, responsible for higher education policy within its borders, took its own approach to promoting (or ignoring) entrepreneurial activity in colleges and universities.

The absence of any kind of national policy direction does not equate to a lack of mobilization around the promotion of EE. Quite the contrary, there are many actors and interest groups engaged in the active promotion of entrepreneurial learning and start-up activity in higher education (Sá & Kretz, 2015). From the bottom up, entrepreneurship experts have advanced EE initiatives following their own professional interests. University administrators have responded to external inducements from philanthropists and provincial economic development programs to create entrepreneurial hubs on their campuses. National and international associations bring together educators and other professionals committed to entrepreneurship, diffusing norms and practices across jurisdictions. Student associations actively organize entrepreneurship-related events and activities. Moreover, student demand has supported the proliferation of entrepreneurship coursework and experiential learning tied to campus incubators and accelerators.

This chapter examines the key actors in the development of EE on Canadian campuses. It explores the role of stakeholders in creating a favorable environment for colleges and universities to become more engaged in supporting entrepreneurship. Then, it reviews how EE has evolved within higher education institutions (HEIs). The conclusion reflects on the nature of the Canadian experience.

C. Sá (✉) · A. Husain
University of Toronto, Toronto, Canada
e-mail: c.sa@utoronto.ca

X. Xu, *Comparative Entrepreneurship Education*,
https://doi.org/10.1007/978-981-99-1835-5_2

2.1 The Influence of Stakeholders

Several stakeholders have actively promoted EE in Canada, directly and indirectly. This section reviews their role in supporting national and provincial entrepreneurial ecosystems, highlighting the participation of different levels of government, and the involvement of actors from the private sector.

2.1.1 Federal Government

In the late 1980s, the federal government began prioritizing entrepreneurship (Industry Canada, 2005; Industry, Science & Technology Canada, 1990; Parsley & Weerasinghe, 2010; Praag & Versloot, 2007; Menna et al., 2016). Echoing global trends, entrepreneurs came to be recognized as key drivers of economic growth and employment (Industry, Science & Technology Canada, 1990; Parsley & Weerasinghe, 2010; Sá & Kretz, 2015). Several studies commissioned by the federal government to examine Canada's global competitiveness and innovation called for enhanced government support for the development of high-tech industry clusters, including proactive investments on specialized educational programs, infrastructure, and strengthening university–industry partnerships (Deloitte Canada, 2017; Industry Canada, 2007; Porter & Martin, 2001). A consensus among policy circles was the pressing need for Canada to create better conditions and incentives for innovative entrepreneurs.

In recent decades, government policy agendas to support innovation included efforts to unleash entrepreneurship in the Canadian economy, including science-based innovation stemming from publicly funded research. Since the 1990s, federal research councils—Canadian Institutes of Health Research, Natural Science and Engineering Research of Canada (NSERC), and Social Sciences and Humanities Research Council of Canada—have increasingly emphasized the commercialization of university inventions and technology transfer (Doern et al., 2016; Rasmussen, 2008; Sá & Litwin, 2011).

Over the past 20 years, the infusion of academic entrepreneurship in federal research support has occurred through dedicated programs such as Networks of Centres of Excellence (Atkinson-Grosjean, 2006), and funding streams within traditional sponsors of academic science emphasizing commercial outcomes (Veletanlić & Sá, 2019). One example of the latter is NSERC's Idea to Innovation (I2I) program. Programs like I2I were established to promote collaboration between academia and industry by supporting partnerships aimed at enhancing industry innovation. Implemented in 2003, this program supports the pre-competitive development of technology and its transfer to industry (NSERC, 2018). While I2I funding represents a small portion of the total funding required to bring an invention to market, the council's intention is to provide targeted support to allow participants to make informed decision as to how to actively pursue a commercialization project, including potentially through a startup. While programs of this kind are couched as a solution to industry's underinvestment in R&D and academia's lack of interest in commercial outcomes, they may be subject to issues that undermine their effectiveness, such as program design, friction among partners, and misaligned incentives at various levels (Veletanlić & Sá, 2019).

Of course, the most far-reaching impact of the federal government in terms of supporting the entrepreneurial ecosystem goes beyond direct support for academic entrepreneurship or EE. Government action affecting entrepreneurial activity encompasses many policy areas such as regulatory affairs, trade, labor market, innovation, regional development, and social policies (Galvão et al., 2020; Parsley & Weerasinghe, 2010; Sá & Kretz, 2015). The federal government contributes to the development of an entrepreneurial ecosystem through the promotion of entrepreneurship via legislation, provision of financing, tax reduction, and creating support infrastructures (Galvão et al., 2020; Parsley & Weerasinghe, 2010).

One example of federal direct support for entrepreneurial activity is its support for the venture capital market, which suffered significantly in the aftermath of the 2008–2009 financial crisis (Deloitte Canada, 2019). The federal government provided direct funding initially through the Business Development Bank of Canada, and then subsequently through the creation of the Venture Capital Action Plan (VCAP) in 2013, which was succeeded by the Venture Capital Catalyst Initiative (VCCI) in 2017. Though the programs were created by different governments—the VCAP under a Conservative government and the VCCI by the succeeding Liberal government—they both aimed at expanding opportunities for early-stage and young companies to access financing and private sector entrepreneurial expertise. These programs have nurtured startups and early-stage businesses, including those on campuses. For example, at the University of Waterloo, McGill University, and the University of Toronto, campus-based and alumni ventures raised CAD $7.36B, $7.06B and $6.57B in venture capital funding respectively from the late 2000s to 2018 (Deloitte Canada, 2019).

2.1.2 Provincial Governments

The growing consensus around the role that HEIs should play in nurturing entrepreneurship has translated into government initiatives toward the development of campus entrepreneurs and startups, with a focus on providing young potential entrepreneurs with the appropriate skills and support (Parsley & Weerasinghe, 2010; Sá & Kretz, 2015). Provincial governments have been active in this space, building upon their longstanding efforts to harness postsecondary institutions to promote regional economic development.

Differences in political and economic contexts as well as social pressures for access have impacted how each province manages higher education across the country (Fisher et al., 2006). Provincial support for EE is based on the premise that through opportunities to learn and engage in entrepreneurship, students can improve their labor market outcomes and contribute to regional economic development. From a policy perspective, stimulating student entrepreneurship is framed as a solution to multiple problems, including youth underemployment, university graduates' lack of workplace skills, job creation, and the innovative performance of local economies (Sá & Kretz, 2015). Thus, provinces take on multiple roles in stimulating entrepreneurship on campuses, such as financier, inducer of entrepreneurial activity, and provider of business support for entrepreneurs involving HEIs as partners. Furthermore, some provinces also hold expectations for HEIs in the identification, development, and enablement of new entrepreneurs.

Provincial government support for EE is more robust in the more populous and economically developed provinces of Ontario, Québec, and British Columbia. These provinces have the greatest number and diversity of campus initiatives as well as the strongest inducements for on-campus EE. In these provinces, governments have implemented mechanisms such as accountability frameworks with HEIs that specifically emphasize the development of entrepreneurial and commercial outcomes. In addition, these provinces have been active in creating provincial agencies tasked with promoting innovation and entrepreneurship.

Over the past 15 years, Ontario has pursued a series of mechanisms to encourage and facilitate on-campus entrepreneurship. Recently, the Ministry of Colleges and Universities promoted this objective through the Strategic Mandate Agreement (SMAs), which are bilateral agreements between the ministry and the province's publicly assisted colleges and universities. SMAs are a key component of the Ontario government's accountability framework for the postsecondary education system (Ministry of Colleges & Universities Ontario, 2020). The current cycle of agreements incentivizes colleges and universities to prioritize labor market outcomes and economic impact. Performance metrics for these objectives are institution-specific and may include requirements for the support or establishment of startup ventures. Under the current cycle of agreements, the University of Toronto, Ryerson University, and York University have requirements pertaining to venture creation. Publicly funded colleges (which mostly provide short-cycle vocational credentials) such as Seneca College and Sheridan College have targets for the creation of startups and for the mentorship of existing community small and medium-size enterprises (SMEs) through their respective on-campus entrepreneurial hubs.

Provincial governments have also established provincial agencies whose remit includes nurturing entrepreneurship on campuses. In Ontario, the Ontario Centre of Innovation, a nonprofit predominantly funded by various ministries, supported the establishment of campus incubators and accelerators between the late 2000s and mid-2010s (Sá & Kretz, 2015). In British Columbia, the provincial agency, Innovate BC, partners with postsecondary institutions, industry, and the National Research Council to encourage the development and commercialization of technologies. Innovate BC currently provides funding opportunities and mentorship for the commercialization of academic research, facilitating partnerships between industry and academia and supporting entrepreneurship. The agency's Innovator Skills Initiative (ISI) supports aspiring technology entrepreneurs on postsecondary campuses. It provides matching grants to technology companies that hire students for internships and work placements. The tech companies must provide students with entrepreneurial and business training as well as mentorship. In its 2021–2022 budget, the Ministry of Jobs, Economic Recovery and Innovation committed CAD $15 million through the ISI (Innovate BC ISI Grant Program, 2018). In Québec, campus entrepreneurship is supported by various ministries such as the Ministry of Economy and Innovation and the Ministry for Small and Medium Enterprises, Regulatory Streamlining and Regional Economic Development. Their initiatives include the Startup Youth program, which has funded 10 entrepreneurship hubs in universities to provide students who are in their last year of college or university an opportunity to launch a scientific startup or to intern with scientific entrepreneurs (Ministry of Economy & Innovation, 2021; Province of Québec, 2017).

Alberta is comparable to British Columbia in terms of GDP, population, and number of HEIs. However, it is an outlier in terms of the generation of direct on-campus entrepreneurial activity compared to the three largest provinces. This is mostly due to the recent rounds of provincial budgets cuts that started in 2019 and have resulted in close to CAD $400 million or about a five-percent decrease per year to the operating budgets of HEIs for 2019 to 2021 (Caldararu et al., 2021; von Scheel, 2021). Though the funding cuts present a major constraint for HEIs for the foreseeable future, the provincial government's recently released Alberta 2030 strategic plan sets a goal to promote the commercialization of academic research, albeit with no specific details on how that would be accomplished (Ministry of Advanced Education, 2021; Omstead, 2021).

Canada's smaller provinces in terms of GDP and population such as Nova Scotia, New Brunswick, Manitoba, and Saskatchewan lack a robust regional entrepreneurial ecosystem. Still, EE programming has evolved in universities. Provincial support for EE seeks to retain university graduates, enhance local business activity, and improve job opportunities. For Nova Scotia's Department of Labour and Advanced Education (LAE), student entrepreneurship is viewed as a means for creating a competitive workforce (Department of Labour and Advanced Education, 2020). LAE invests in a postsecondary education sandbox program to strengthen early-stage innovation on campuses (Department of Labour and Advanced Education, 2019). This program brings students together with mentors and advisers to help them develop entrepreneurial skills. The intention is to benefit employers and to create jobs for graduates. There are 10 sandboxes on university and college campuses that receive funding for activities and programming. In the fall semester of the 2019–2020 academic year, sandboxes developed and delivered programming that reportedly served 8,471 students who participated in 237 activities (Department of Labour & Advanced Education, 2020).

Similarly, the New Brunswick government has sought to develop more startups and increase university research as part of its innovation agenda (Department of Post-Secondary Education, Training and Labour, 2020). The New Brunswick Innovation Foundation (NBIF) supports entrepreneurship and the commercialization of research through programs such as the Startup Investment Fund, Research Innovation Fund, and the Innovation Voucher Fund (NBIF, 2021). It also sponsors a startup competition,

Breakthru, to encourage the development of entrepreneurial business ventures both on and off campuses. Participants are provided access to startup capital, mentors, local entrepreneurs, and industry leaders through bootcamps.

2.1.3 Private Sector

In addition to governments, stakeholders from the private sector such as successful entrepreneurs, industry, nonprofit organizations, alumni, and private foundations play an important role in EE (Bischoff et al., 2018; Kretz & Sá, 2015; Sá et al., 2014). These external groups join forces with HEIs to interact with government agencies, funding bodies, and financial institutions to create capacity for supporting campus entrepreneurs (Fetters et al., 2010; Regele & Neck, 2012). Their support has led to the development of support services, incubators, accelerators, entrepreneurship centres, and technology parks in HEIs.

Institutions that have links with established entrepreneurial ecosystems and networks of entrepreneurial supportive organizations are able to provide more diverse and targeted programming and services. Universities usually capitalize on these relationships with private sponsors and donors to create and expand their EE infrastructure and programming.

At Simon Fraser University in Vancouver, British Columbia, multiple stakeholder groups including various levels of government, nonprofits, and alumni support the Coast Capital Savings Venture Connection and the Chang Institute for Entrepreneurship. The latter offers activities and courses, residences for graduate students, and incubation facilities. It is supported and funded by Charles Chang, a business alumni, entrepreneur and venture capitalist, as well as other public and private sponsors (Simon Fraser University, n.d.). Students, faculty, and alumni can participate in extracurricular programming at Coast Capital Savings Venture Connection, which is the longest-running entrepreneurship program at the university. It supports aspiring entrepreneurs through founder training, mentorship, and entrepreneurship co-op opportunities, as well as by providing industry connections, office space, startup services for idea validation, and support for venture development.

At Toronto Metropolitan University, Ontario, several EE initiatives have flourished with the support of public and private sponsors. DMZ, the university's flagship incubator launched in 2009, is supported by national and international businesses and other organizations. It provides students, alumni, and young entrepreneurs with access to support, mentorship, sandbox programming, workshops, and access to seed funding. Student entrepreneurs can partake in a sandbox program that provides EE programming, workshops, access to mentors and funding that are all designed to provide experiential learning, facilitate innovative ideas and help launch student ventures (Toronto Metropolitan University, n.d.).

2.2 EE in Colleges and Universities

EE programming has grown and diversified as HEIs provide curricular, cocurricular, and extracurricular entrepreneurship offerings to serve more students outside of business schools (Duval-Couetil, 2013; Greene & Storey, 2004; Kuratko, 2005; Rideout & Gray, 2013; Sá & Kretz, 2015; Toscher, 2019). Colleges and universities are actively involved in offering EE and providing a network of practitioners with the necessary facilities to support students, faculty, and staff interested in entrepreneurship (Parsley & Weerasinghe, 2010).

Universities that have larger student populations, are in larger economic regions, and are more closely linked to established entrepreneurial ecosystems tend to have more expansive EE offerings. Most institutions, small and large, rely on entrepreneurship centres to offer extracurricular and experiential learning activities. These units usually run pitch competitions, informational webinars, community events, mentorship and coaching, and provide access to seed funding. In terms of their orientation, cocurricular programs may focus on introducing students across disciplines to entrepreneurship or emphasize systematically designed opportunities for students to gain and refine technical and business skills (Sá & Kretz, 2015).

While entrepreneurship hubs are more commonly linked to business and engineering schools, entrepreneurship-related courses can be found across academic fields such as the arts, social sciences, and health sciences. Topics include business plans and ventures, skill development courses such as marketing, management, and entrepreneurial thinking, as well as specialized courses in areas like family business management, social entrepreneurship, and innovation.

Located in Montreal, Québec—Canada's second-largest city and with one of the largest entrepreneurial ecosystems—McGill University offers a wide variety of entrepreneurship programming. Students from various disciplines have the opportunity to take major and minor concentrations in entrepreneurship through the Desautels Faculty of Management. The Dobson Centre of Entrepreneurship offers faculty, students, and alumni a variety of services, mentorship opportunities, and resources to support the growth of startups (McGill University, n.d.).

In smaller institutions, EE programming may be spearheaded by a centrally supported entity. Institutions such as the University of Manitoba offer some curricular programming through their entrepreneurship center. Curricular programming has been expanded outside of business schools to other faculties such as the arts, engineering science, and agricultural science. On-campus entrepreneurship centres offer services such as business plan and venture creation competitions, webinars, online venture coaching, and connections to community entrepreneurial mentors. However, smaller student populations, which lack provincial funding and direct support, and smaller regional economies limit the range of opportunities for students.

College EE offerings also tend to be somewhat more circumscribed as compared to those of larger universities. This may be because of their lack of integration into the entrepreneurial ecosystem focused on science-based technology commercialization, which is centered in universities. However, the colleges' traditional mission to provide vocational training and support community development underpins the work of their entrepreneurship hubs. Toronto's Centennial College, for example, offers EE courses for students in a wide variety of fields such as business, hospitality and tourism, and media studies. The college's Centre of Entrepreneurship provides experiential learning and business advising programs to assist students and broader community members launch or grow their business ventures (Centennial College, n.d.).

2.2.1 Entrepreneurship Hubs

Entrepreneurship hubs such as entrepreneurship centres, incubators, or accelerators tend to provide a focal point for a variety of cocurricular activities in colleges and universities. They offer students access to counselling and mentoring services, networking events, workshops, and speaker series. Entrepreneurs-in-residence provide a link to the external entrepreneurial ecosystem for students and aspiring entrepreneurs (Sá & Kretz, 2015).

The number of entrepreneurship hubs has grown considerably since the late 1970s when Canada's first entrepreneurship center was created (Menzies, 2009; Sá, 2018; Sá et al., 2014). Entrepreneurship centres have also increased in importance as they tend to be conduits between students, entrepreneurs, local business owners, and local community support organizations. The breadth of programming and services of entrepreneurship centres varies from institution to institution and is largely influenced by the proximity of the institution to larger economic regions and the size of student populations. This variance is also influenced by the availability of funding and support from the stakeholders described above.

Thus, in smaller universities such as St. Mary's University in Halifax, Nova Scotia, the entrepreneurship centre in the business school focuses on providing students with opportunities to connect with local businesses for internship, co-op, and consulting activities. Universities such as Toronto, British Columbia, and Waterloo have a more dispersed set of entrepreneurship hubs playing several roles, from helping students "ideate" new business to launching successful bids for private investment funding for up-and-running startups.

At York University's Lassonde School of Engineering, the Bergeron Centre of Engineering Excellence offers the Bergeron Entrepreneurs in Science and Technology (BEST) program. It provides a wide variety of entrepreneurship programming and initiatives designed for engineering students who are looking to start their own ventures or develop entrepreneurial skills. Programs and offerings include the BEST Certificate, which integrates multidisciplinary entrepreneurial courses from different faculties (business, law, design, engineering). Also available are co-op opportunities, workshops, pitch competitions, hackathons, and mentorship. The programming at the Bergeron Centre is facilitated by a diverse group of stakeholders such as former alumni, business community members, and entrepreneurship support organizations.

On college campuses, entrepreneurship centres target not just students but also community members who are interested in entrepreneurship (Centennial College, n.d.; Humber College, 2018; Seneca College, n.d.). Programming similarly consists of workshops, competitions, bootcamps, mentorship, and events. On some college campuses there is curricular EE programming is integrated and coordinated with programs and services offered by the institution's entrepreneurship centres. Centennial College's Centre of Entrepreneurship, an on-campus entrepreneurship center, has developed a variety of extracurricular programming that includes mentoring, workshops, bootcamps as well as market-readiness and incubation programs for both students and community members looking to launch or develop a business. Some of the offerings are presented in experiential learning formats. The centre is also integrated with the curriculum offered by the college through its work integrated learning option where entrepreneurship students can participate in any of the programs such as co-op, internships or field placements (Centennial College, n.d.).

Accelerators and incubators are important facets of entrepreneurial ecosystems. They facilitate entrepreneurship on campuses by offering support, services, and seed funding to early-stage and startup campus entrepreneurs. While accelerators focus on nurturing more advanced ventures, incubators focus on early-stage ventures and the development of fundamental entrepreneurial skills such as business development, incorporation, shareholder agreements, or human resource management. Characteristics of accelerators and incubators vary widely across institutions; they offer a combination of services such as access to mentors, office space, workspaces, business services, structured programming, and workshops (Sá et al., 2014).

The University of Toronto supports entrepreneurship across its three campuses through about a dozen entrepreneurship hubs. A robust support infrastructure has emerged organically from interested faculty and administrators from different units, as well as in response to external opportunities. The Creative Destruction Lab (CDL) is an internationally renowned accelerator based in the Rotman School of Management. Founded in 2012, the CDL has since expanded domestically and globally into many other campuses including the University of British Columbia, University of Calgary, HEC Montréal, Dalhousie University, and institutions abroad. As an elite accelerator program focused on high-growth scale-up firms, the CDL provides founders with coaching, relationship-building assistance, strategic guidance, in addition to a range of business support services. MBA students at the Rotman School of Management can participate in the CDL Fellowship Program, which enrols 20 students every year. Demonstrating a commitment to entrepreneurship and venture creation, the students are provided with funding, support, extracurricular courses, and access to events. All these enhance and develop their startup experience (University of Toronto, n.d.).

The University of Waterloo also supports on-campus entrepreneurship from early stage to market launch through campus incubators and accelerators. Waterloo has reportedly supported the creation of 314 companies by university graduates and has produced the largest number of venture capital-backed entrepreneurs in Canada between 2006 and 2018 (Deloitte Canada, 2019). Founded in 2008, Velocity is a pre-seed incubator that provides students and community members from the wider

Waterloo region with programming such as bootcamps, mentorship, as well as access to tools, technologies, labs, workspaces, grants, and funding. One of Velocity's goals is to facilitate greater integration between industry, the business community, and entrepreneurial students. Velocity has also developed a pre-incubator program, Concept, specifically for students who are looking for extracurricular experiential EE to complement their studies. Students can partake in workshops, benefit from mentoring and coaching, and access funding opportunities to develop their entrepreneurial ideas.

2.3 Conclusion

Many actors contribute to EE across Canadian HEIs. Relevant stakeholders include various levels of government, regional economic development organizations, successful entrepreneurs, alumni, nonprofit organizations, private foundations, and alumni. On-campus programming has evolved organically over the past two decades, as faculty and administrators pursue their professional goals, respond to incentives provided by public and private sponsors, and meet the demand of students for opportunities to engage in entrepreneurship.

Larger universities located in or close to densely populated cities in more economically robust provinces have led the way in terms of entrepreneurship programming, deploying greater resources to support on-campus entrepreneurs. They tend to support high-growth, innovative startups that can scale up. As such, there is a nexus between the activities of entrepreneurship educators and the commercial engagement of academic entrepreneurs seeking to bring inventions to market. While the latter may not include launching or working with a startup, these two streams of entrepreneurial activity are both technology-driven and research-based.

Colleges also play an important role in EE, as it meets their traditional mission of facilitating successful labor market transitions for students and supporting local business. Their niche in the EE ecosystem lies in helping the students they serve—who are disproportionally from the lower socioeconomic group, immigrant, and minorities—gain relevant skills to launch small businesses and gain useful skills. There is some integration between the curricular offerings by the college and extracurricular programming options available from the institutions' entrepreneurship centers, specifically through work integrated learning opportunities for students and pitch competitions. These activities tend not to be as lavishly supported as compared to those of larger university hubs, as they are not seen as part of the broader technology-driven agenda to nurture high-growth startups.

Overall, the extent of institutional commitments to EE vary according to size, location, and institutional strengths. Proximity to established entrepreneurial ecosystems and networks of supportive organizations does seem to expand the possibilities for campus programming. Firstly, it allows institutions to provide more diverse and targeted programming, services, and support for entrepreneurs on their campuses. These take on the form of curricular offerings such as degree and diploma programs, certificates, as well as cocurricular and extracurricular offerings. In addition, while entrepreneurship centres and programs typically reside in business schools and engineering schools, campus-wide activities are increasingly available at many institutions. Secondly, it facilitates the emergence and development of on-campus entrepreneurship hubs such as centres, incubators, and accelerators, all of which are instrumental in the provision of cocurricular and extracurricular EE. Finally, programming diversity and the emergence of on-campus entrepreneurship hubs have cascaded to move away from the traditional practice of teaching entrepreneurship through theoretical and case-based methods to project-based and experiential learning methods.

References

Alberta Advanced Education. (2021, April). *Alberta 2030: Building skills for jobs.* Retrieved from https://open.alberta.ca/dataset/24e31942-e84b-4298-a82c-713b0a272604/resource/b5a2072e-8872-45f9-b84d-784d0e98c732/download/ae-alberta-2030-building-skills-for-jobs-10-year-strategy-post-secondary-education-2021-04.pdf

Atkinson-Grosjean, J. (2006). *Public science, private interests: Culture and commerce in Canada's Networks of Centres of Excellence.* University of Toronto Press. Retrieved from https://doi.org/10.3138/9781442678897

Bischoff, K., Volkmann, C. K., & Audretsch, D. B. (2018). Stakeholder collaboration in entrepreneurship education: An analysis of the entrepreneurial ecosystems of European higher educational institutions. *The Journal of Technology Transfer, 43*(1), 20–46.

Caldararu, A., Howes, B., Marcotte, P., & Wilson, G. (2021, March 10). Opinion: Post-secondary cuts a 'circuit breaker' for Alberta's economy. *Edmonton Journal.* Retrieved from https://edmontonjournal.com/opinion/columnists/opinion-post-secondary-cuts-are-a-circuit-breaker-for-albertas-economy

Centennial College. (n.d.). *Centre of entrepreneurship.* Retrieved from https://www.centennialcollege.ca/about-centennial/centres-and-institutes/applied-research-innovation-and-entrepreneurship-services/research-and-innovation-centres/centre-of-entrepreneurship/

Deloitte Canada. (2017). *Canada at a new crossroads: 25 years later.* Retrieved from https://www2.deloitte.com/content/dam/Deloitte/ca/Documents/insights-and-issues/ca-en-insights-issues-canada-at-a-new-crossroads.PDF

Deloitte Canada. (2019, November). *University of Waterloo economic impact assessment.* Retrieved from https://uwaterloo.ca/about/sites/default/files/uploads/documents/university_of_waterloo_economic_contribution_analysis_2019.pdf

Department of Labour and Advanced Education. (2019, March). *Business plan 2019–2020: Department of labour and advanced education.* Retrieved from https://novascotia.ca/government/accountability/2019-2020/2019-2020-business-plan-Department-of-Labour-and-Advanced-Education.pdf

Department of Labour and Advanced Education. (2020, August). *Accountability report 2019–2020: Department of labour and advanced education.* Retrieved from https://beta.novascotia.ca/sites/default/files/documents/1-2532/accountability-report-2019-20-department-labour-and-advanced-education-en.pdf

Doern, G. B., Castle, D., & Phillips, P. W. (2016). *Canadian science, technology, and innovation policy: The innovation economy and society nexus.* McGill-Queen's University Press—MQUP.

Duval-Couetil, N. (2013). Assessing the impact of entrepreneurship education programs: Challenges and approaches. *Journal of Small Business Management, 51*(3), 394–409.

Fetters, M. L., Greene, P. G., Rice, M. P., & Butler, J. S. (2010). *The development of university-based entrepreneurship ecosystems: Global practices*. Edward Elgar Publishing.

Fisher, D., Rubenson, K., Bernatchez, J., Clift, R., Jones, G., Lee, J., MacIvor, M., Meredith, J., Shanahan, T. and Trottier, C., 2006. *Canadian federal policy and postsecondary education*. Centre for Policy Studies in Higher Education and Training, Fac. of Education, University of British Columbia.

Galvão, A. R., Marques, C. S. E., Ferreira, J. J., & Braga, V. (2020). Stakeholders' role in entrepreneurship education and training programmes with impacts on regional development. *Journal of Rural Studies, 74*, 169–179.

Government of New Brunswick. (2020, November). *Post-secondary education, training and labour annual report 2019–2020*. Retrieved from https://www2.gnb.ca/content/dam/gnb/Departments/petl-epft/PDF/Publications/annual-report-2019-2020.pdf

Greene, F. J., & Storey, D. J. (2004). An assessment of a venture creation programme: The case of Shell Live WIRE. *Entrepreneurship & Regional Development, 16*(2), 145–159.

Humber College. (2018, August). *The centre for entrepreneurship*. Retrieved from https://humber.ca/staff/announcement/centre-entrepreneurship-cfe

Industry Canada. (2005). *Entrepreneurship and small business policy in Canada*.

Industry Canada. (2007). *Mobilizing science and technology to Canada's advantage*. Retrieved from https://ised-isde.canada.ca/site/plans-reports/sites/default/files/attachments/SandTstrategy.pdf

Industry, Science and Technology Canada. (1990). *National forum on entrepreneurship: The federal government's response to the recommendations*. Retrieved from https://publications.gc.ca/collections/collection_2019/isde-ised/c28/C28-12-1990-eng.pdf

Innovate BC ISI Grant Program. (2018, October). *Innovate BC innovator skills initiative program guidelines*. Retrieved from https://www.newventuresbc.com/wp-content/uploads/2019/11/Innovate-BC-ISI-Program-Guidelines-Public-Nov-20-2019.pdf

Kretz, A., & Sá, C. M. (2015). Students and startups: How new forms of entrepreneurial learning and practice redraw university boundaries. In E. Reale & E. Primeri (Eds.), *The transformation of university institutional and organizational boundaries* (pp. 83–105). Sense Publishers.

Kuratko, D. F. (2005). The emergence of entrepreneurship education: Development, trends, and challenges. *Entrepreneurship Theory and Practice, 29*(5), 577–598.

McGill University. (n.d.). *Desautels programs*. Retrieved from https://www.mcgill.ca/desautels/programs/bcom/academics/minors-non-management-students/entrepreneurship-minors-non-management-students

Menna, A., Catalfamo, H., & Girolamo, C. (2016). Applied entrepreneurship policy: Ontario's colleges in the age of globalization. *Educational Planning, 23*(2), 59–74.

Menzies, T. V. (2009). *Entrepreneurship and the Canadian universities: Strategies and best practices of entrepreneurship centres*. Faculty of Business, Brock University.

Ministry of Colleges and Universities Ontario. (2020). *College and university strategic mandate agreements 2020–2025*. Retrieved from https://www.ontario.ca/page/college-and-university-strategic-mandate-agreements-2020-2025

Ministry of Economy and Innovation. (2021, September 14). *Startup Québec* [Startup Québec]. Retrieved from https://www.economie.gouv.qc.ca/bibliotheques/programmes/aide-financiere/programme-daide-a-lentrepreneuriat-paen/soutien-au-fonctionnement-et-a-des-projets-structurants-des-organismes-en-entrepreneuriat/startup-quebec/

Natural Sciences and Engineering Research Council of Canada. (2018, March). *Final report evaluation of commercialization of research: Idea to innovation grants*. Retrieved from https://www.nserc-crsng.gc.ca/_doc/Reports-Rapports/Evaluations/2018/CoR121Final_e.pdf

New Brunswick Innovation Foundation (NBIF). (2021). *Impact report 2019–2020*. Retrieved from https://nbif.ca/wp-content/uploads/2021/01/NBIF_Annual-Report_ENG-Web.pdf

Omstead, J. (2021, April 30). Commercial research and jobs focus of Alberta's 10-year post-secondary plan. *CBC News*. Retrieved from https://www.cbc.ca/news/ige/igey/alberta-2030-plan-advanced-education-1.6007695

Parsley, C., & Weerasinghe, M. (2010, December). *The teaching and practice of entrepreneurship within Canadian higher education institutions*. Industry Canada. Retrieved from https://publications.gc.ca/collections/collection_2011/ic/Iu188-103-2010-eng.pdf

Porter, M. E., & Martin, R. L. (2001). *The competitive advantage of nations*. Retrieved from http://www-2.rotman.utoronto.ca/rogermartin/Competitiveness.pdf

Praag, M. V., & Versloot, P. H. (2007). What is the value of entrepreneurship? A review of recent research. *Small Business Economics, 29*(4), 351–382.

Province of Québec. (2017). *Québec research and innovation strategy*. Retrieved from https://www.ryerson.ca/content/dam/provost/PDFs/RyersonUniversity-AcademicPlan2020-25.pdf

Rasmussen, E. (2008). Government instruments to support the commercialization of university research: Lessons from Canada. *Technovation, 28*(8), 506–517.

Regele, M. D., & Neck, H. M. (2012). The entrepreneurship education subecosystem in the United States: Opportunities to increase entrepreneurial activity. *Journal of Business and Entrepreneurship, 23*(2), 25–47.

Rideout, E. C., & Gray, D. O. (2013). Does entrepreneurship education really work? A review and methodological critique of the empirical literature on the effects of university-based entrepreneurship education. *Journal of Small Business Management, 51*(3), 329–351.

Sá, C. (2018). What does success look like for entrepreneurship education? *International Journal of Chinese Education, 7*(2), 193–206.

Sá, C., & Litwin, J. (2011). University-industry research collaborations in Canada: The role of federal policy instruments. *Science and Public Policy, 38*(6), 425–435.

Sá, C. M., Kretz, A., & Sigurdson, K. (2014). *The state of entrepreneurship education in Ontario's colleges and universities*. Higher Education Quality Council of Ontario.

Sá, C. M., & Kretz, A. J. (2015). *The entrepreneurship movement and the university*. Palgrave MacMillan.

Seneca College. (n.d.). *Seneca innovation: Helix*. Retrieved from https://www.senecacollege.ca/innovation/helix.html

Simon Fraser University. (n.d.). *About the Charles institute for entrepreneurship*. Retrieved from https://www.sfu.ca/changinstitute/about/

Toronto Metropolitan University. (2020). *Academic plan 2020–2025*. Retrieved from https://www.torontomu.ca/content/dam/provost/PDFs/2020-25_academic_plan_final_web_FINAL-s.pdf

Toronto Metropolitan University. (n.d.). *About DMZ*. Retrieved from https://dmz.ryerson.ca/about

Toscher, B. (2019). Entrepreneurial learning in arts entrepreneurship education: A conceptual framework. *Artivate, 8*(1), 3–22.

University of Toronto. (n.d.). *Rotman: Creative destruction lab fellowship program*. Retrieved from https://www.rotman.utoronto.ca/Degrees/MastersPrograms/MBAPrograms/FullTimeMBA/Program/CDLFellowship

Veletanlić, E., & Sá, C. (2019). Government programs for university-industry partnerships: Logics, design, and implications for academic science. *Research Evaluation, 28*(2), 109–122.

Von Scheel, E. (2021, March 22). Big cuts hinder ability to help economic recovery, post-secondary groups say. *CBC News*. Retrieved from https://www.cbc.ca/news/ige/igey/cuts-budget-2021-alberta-university-college-1.5955547

Chapter 3
Entrepreneurship Education in China

Xiaozhou Xu and Weihui Mei

Entrepreneurship education (EE) in China can be traced to 1998 when Tsinghua University held the country's first Entrepreneurship Plan Competition (Wang, 2016). In the past two decades, the government, higher education institutions (HEIs), enterprises and other stakeholders have played different roles in the development of China's EE system. This chapter examines the state of EE in China, its characteristics, and future directions.

3.1 Context

3.1.1 Unemployment Pressure of University Graduates

At the end of the 1990s, China implemented the higher education expansion policy. Since then, enrollments and graduates have increased dramatically (see Fig. 3.1). From 2001 to 2019, the gross enrollment rate of higher education increased from 13.3 to 51.6%, and the number of HEI graduates rose from approximately 1.04 million to 7.59 million (Ministry of Education of the People's Republic of China, 2020). As many more university graduates enter the market, one of the main challenges for China is to utilize the potential of new graduates more productively in the economy (ILO, 2015). The continuous rise of HEI graduates and the annual unemployment rate of 20–30 percent have increased the unemployment pressure among college students (Feng, 2013, p. 19). Therefore, the idea of "promoting employment through entrepreneurship" gradually entered the vision of decisionmakers, and has become an important policy agenda for HEIs and the government for dealing with unemployment (Zhang & Su, 2016).

X. Xu (✉) · W. Mei
College of Education, Zhejiang University, Hangzhou, China
e-mail: xxiaozhou@163.com

X. Xu, *Comparative Entrepreneurship Education*,
https://doi.org/10.1007/978-981-99-1835-5_3

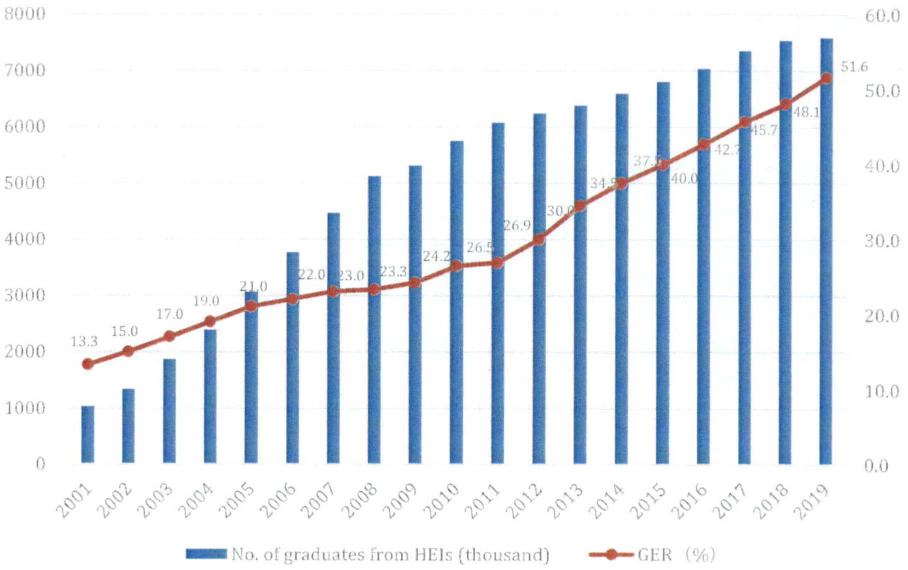

Fig. 3.1 University graduates and gross enrollment rate in China, 2001–2019. *Source* Ministry of Education (2020). Statistical Bulletin of National Education Development 2001–2019. http://www.moe.gov.cn/jyb_sjzl/sjzl_fztjgb/

3.1.2 Shortage of Skilled and Innovative Talents

In China, university graduates' ability cannot meet the needs of economic transformation and the Fourth Industrial Revolution. The country is going through a dramatic economic transformation, and the next 10 years will be a crucial period for China to realize its "Chinese Dream." Presently, the country focuses on and prioritizes knowledge innovation and innovation-based entrepreneurship (Mei & Symaco, 2022). However, the emergence of the Fourth Industrial Revolution has put forward new requirements for youth competences. Based on a global employer survey, the World Economic Forum (WEF) pointed out in 2018 that analytical thinking and innovation as well as active learning and learning strategies were skills that would continue to grow in prominence by 2022. Besides, "'human' skills such as creativity, originality and initiative, critical thinking, persuasion, and negotiation will retain or increase their value" (WEF, 2018, p. 12). Meanwhile, the European Commission identified entrepreneurship competence as one of the eight key competences for lifelong learning (European Commission, 2018). Therefore, schools and HEIs must prepare the new generation "for jobs that have not yet been created, for technologies that have not yet been invented, to solve problems that have not yet been anticipated" (OECD, 2018, p. 2). However, the Chinese education system pays too much attention to knowledge while ignoring the promotion of curiosity and imagination. It is too eager for quick success and too utilitarian in its value orientation (Qian, 2017).

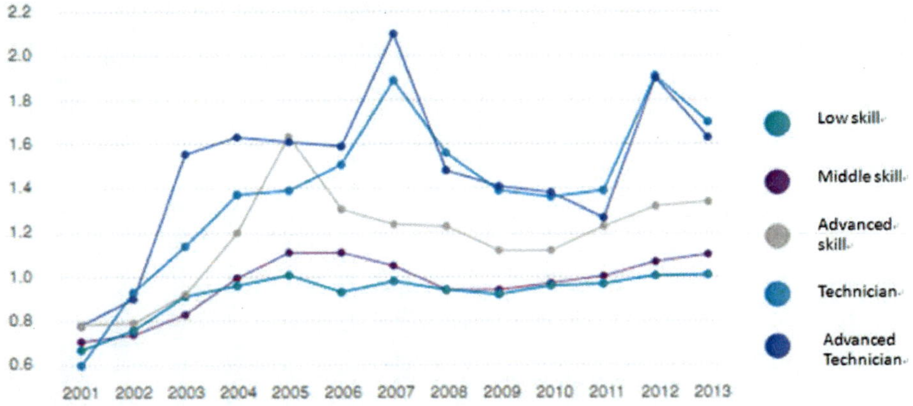

Fig. 3.2 Ratio of job demand to job seekers. *Source* Research on the Skills Gap in China's Labor Market (2016, p. 93)

Therefore, the introduction of innovation and EE has been gradually regarded as a tremendous breakthrough and opportunity to reform the entire education system in China, rather than only cultivating college students to start their own business.

Though the higher education system has expanded dramatically, there is still a serious shortage of skilled and innovative talents. By the end of 2016, there were 165 million skilled laborers nationwide, which accounted for 21.3% of the total employed population; and 47.91 million highly skilled workers, which only accounted for 6.2% of the total employed population (China Labor Security Newspaper, 2018). According to *Research on the Skills Gap in China's Labor Market* (Tsinghua University & Fudan University, 2016), the gap is much greater for the higher-skilled workforce. As shown in Fig. 3.2, the demand for advanced technicians is the most urgent: the ratio of demand to job seekers has been higher than 1.4 since 2003, and reached a high of almost 2.2 in 2007. Meanwhile, the demand and supply of low- and middle-skilled workers have reached a balance.

3.1.3 Low Level of Technology Transfer and Rate of Entrepreneurship

Though R&D investment has continuously increased in China, the level of technology transfer and rate of entrepreneurship are still quite low due to the lack of entrepreneurship culture. According to the statistics by the Ministry of Science and Technology, R&D funding grew dramatically from RMB300.3 billion in 2006 to RMB2214.4 billion in 2019, and the ratio of R&D funding to GDP grew from 1.39% in 2006 to 2.23% in 2019 (MOST, 2020).A comprehensive survey of Chinese college graduates' employment shows that college graduates' entrepreneurship rate rose from

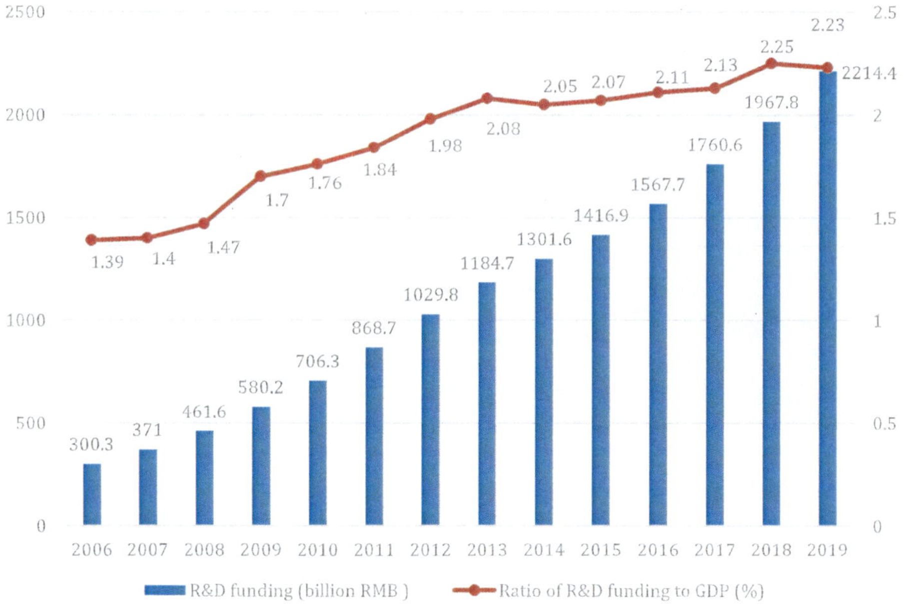

Fig. 3.3 Amount of R&D funding and ratio of R&D funding to GDP. *Source* MOST (2020)

1% in 2008 to 2.3% in 2013 (Mycos Institute, 2010, 2014), but this is still far behind that of other age groups and of other countries (Fig. 3.3).

3.2 EE Policies in China

The Chinese government at various levels has issued many policies supporting entrepreneurship and EE (see Table 3.1). The Mass Entrepreneurship and Innovation agenda introduced in 2014 highlights the role of entrepreneurship for social and economic development in China, in addition to various policies that further support EE. In 2010, the Ministry of Education (MOE) issued a circular promoting EE in HEIs and the self-employment of university graduates. This circular pushes the development of entrepreneurship courses and the recruitment of qualified entrepreneurship faculty members. In 2012, the MOE required all HEIs to provide two-credit introductory entrepreneurship courses for students. *Basic Requirements of Teaching Entrepreneurship Education in Higher Education Institutions* was introduced in the same year, which provided guidelines on teaching objectives, principles, content, and methods of the 'Basics of Entrepreneurship' course (MOE, 2012; Mei & Symaco, 2022). The State Council also issued a circular in 2015 promoting additional reform and advancement of EE in HEIs, including the need to reform teaching and assessment methods in EE, promote better EE practices, and providing better financial support for the program (State Council of the People's Republic of China, 2015).

Table 3.1 Main EE policies in China

Year	Policy	Policymaker	Main points
2003	SME Promotion Law of the People's Republic of China	NPC	Support SMEs' entrepreneurship and innovation, promoting the development of SMEs and expanding employment
2004	Opinions on Implementing Chinese Youth Entrepreneurship Action to Promote Youth Employment	CYLCC, MOHRSS	Support Chinese youth entrepreneurship by helping 50,000 young people start their own business each year
2006	Identification and Regulation of High-tech Business Incubators	MOST	Facilitate the commercialization of research findings
2007	Regulation of Venture Capital Guidance Funds for High-tech SMEs	MOF, MOST	Provide venture capital for high-tech SMEs
2008	Employment Promotion Law of the People's Republic of China	NPC	A specialized law supporting employment and entrepreneurship
2008	Guidance Opinions on Promoting Entrepreneurship Driving Employment	State Council	Implement the development strategy of expanding employment and facilitate employment by entrepreneurship
2010	Identification of High-tech Entrepreneurship Internship Base for College Students	MOE, MOST	Provide comprehensive service platform for college students to practice and train for entrepreneurship and employment
2010–2012	Notice on Implementing College Students Entrepreneurship Leading Plan	Nine Departments & Ministries	Help 450,000 college students to start their own businesses
2012	Notice on Supporting Development of SMEs	State Council	Help SMEs overcome difficulties
2012	Basic Requirements of Teaching Entrepreneurship Education in Higher Education Institutions	MOE	Set the core course "Basics of Entrepreneurship"

(continued)

Table 3.1 (continued)

Year	Policy	Policymaker	Main points
2014	Notice on Supporting the Development of Venture Capital Investment Institutions	NDRC	To facilitate the development of venture capital investment industry
2014–2017	Notice on Implementing College Students Entrepreneurship Leading Plan	Nine Departments & Ministries	Help 800,000 college students to start their own businesses
2015	Opinions on Deepening System and Mechanism Reform and Implementing Innovation Driving Development Strategy	State Council	Implement innovation-driven strategy
2015	Implementation Opinions on Deepening the Reform of Innovation and Entrepreneurship Education in Higher Education Institutions	State Council	Comprehensive guidance on Innovation and entrepreneurship education in higher education institutions
2021	Guidance on Further Supporting College Students' Innovation and Entrepreneurship	State Council	Comprehensive guidance for higher education institutions to support students to implement entrepreneurial activities

Note CYLCC: Central Committee of the Communist Youth League; MOE: Ministry of Education; MOF: Ministry of Finance; MOHRSS: Ministry of Human Resources and Social Security; MOST: Ministry of Science and Technology; NDRC: National Development and Reform Commission; NPC: National People's Congress

Lundström and Stevenson (2005) classified entrepreneurship policies into six areas: promotion of entrepreneurship; entrepreneurship in the education system; environment for startups; access to startup, seed, and early-stage financing; business startup support; and targeting of underrepresented groups and tech-starters. EE policies are a part of entrepreneurship policies. This chapter classifies EE policies into three categories: education about entrepreneurship, education for entrepreneurship, and education through entrepreneurship.

3.2.1 "Education About Entrepreneurship" Policy in China

In 2005, the International Labor Organization, in partnership with the All-China Students' Federation, launched the project called Know about Business (KAB) Entrepreneurship Education Program (China), which aimed to raise entrepreneurship awareness and improve the entrepreneurial competence of undergraduate students. By February 2021, the KAB project had trained more than 11,300 teachers from over 1,840 colleges and universities, and established the KAB entrepreneurship club in more than 400 colleges and universities. Over two million college students have participated in the KAB (2011). The KAB course is the first "Education about Entrepreneurship" elective offered to college students in China.

In 2012, the MOE issued the official document *Basic Requirements of Teaching Entrepreneurship Education in Higher Education Institutions*, which indicated the teaching objectives, principles, content, methods, and organization of the "Basics of Entrepreneurship" course. All HEIs were required to provide this two-credit course to raise students' entrepreneurial awareness and spirit.

3.2.2 "Education for Entrepreneurship" Policy in China

In 2015, the State of Council of China published the document *Implementation Opinions on Deepening the Reform of Innovation and Entrepreneurship Education in Higher Education Institutions*, which put forward three general objectives and nine main tasks and measures of EE. The central aim of this document was to cultivate innovative and entrepreneurial competencies of college students. EE in China gradually gained legitimacy and became increasingly important. The general objective was: "By 2020, innovation and entrepreneurship education system should be established. The quality of talent training has been greatly improved, and the students' competencies of innovation and entrepreneurship have been obviously enhanced" (State Council of the People's Republic of China, 2015).

The nine main tasks and measures of EE were: (1) Improving the quality standard of cultivating talents; (2) Innovating the talent cultivation mechanism; (3) Improving the innovation and EE curriculum system; (4) Reforming teaching and assessment approaches; (5) Reinforcing entrepreneurial practice; (6) Reforming the administration system of entrepreneurship teaching and students' status; (7) Improving the faculty's capacity building of teaching innovation and EE; (8) Ameliorating entrepreneurship guidance service; and (9) Improving the financial support and policy guarantee system for innovation and entrepreneurship (State Council of the People's Republic of China, 2015).

3.2.3 "Education Through Entrepreneurship" Policy in China

Government at all levels in China have also issued and enacted many policies to support graduate entrepreneurship, including financial support, business startup support, and reducing barriers to entry.

In terms of financial support, in 2003, the Office of State Council issued the *Notice on Boosting the Employment of the Graduate of Higher Education Institutions*, which encouraged the local government to provide microfinance and guarantees for college graduates. By 2018, every province in China provided a loan of RMB50,000 to RMB500,000 with low interest rates. The central and provincial governments also offered a variety of entrepreneurship and investment funds to encourage and support graduate entrepreneurship. These included the *National Innovation Fund for Hi-tech SMEs*, *Entrepreneurial Investment Guidance Fund for Hi-tech SMEs*, *Entrepreneurship Fund for College Graduates*, and *Special Fund for SMEs*. Besides government funding, college students and graduates could also seek funds from venture capital and angel investments.

China's central and provincial governments have also established organizations to support graduate entrepreneurship, including internship bases, entrepreneurship spaces, and incubators. The staff in these organizations offer college students mentorship and counseling services. In 2010, for example, the MOE and the Ministry of Science and Technology issued the document, *Identification of High-tech Entrepreneurship Internship Base for College Students*, which stated that the University Science and Technology Park and Hi-tech Industrial Development Zone should work as entrepreneurship internship bases for college students. By 2018, more than 100 entrepreneurship internship bases in the University Science and Technology Park were identified. In March 2015, the General Office of the State Council set out the *Opinions on Developing Entrepreneurship Space to Advance Mass Innovation and Entrepreneurship*, which proposed to construct low-cost and shared entrepreneurship spaces to encourage science and technology personnel and college students to start businesses. In October 2021, the General Office of State of Council of China issued *Guidance on Further Supporting College Students' Innovation and Entrepreneurship*, which emphasizes not only the improvement of college students' innovation and entrepreneurship ability, but also the systematic support for college students' entrepreneurial practices. These supports include optimizing the innovation and entrepreneurship environment; strengthening the construction of innovation and entrepreneurship service platform; promoting the implementation of financial and tax support policies for college students' innovation and entrepreneurship; promoting technology transfer; and strengthening the information service for college students' innovation and entrepreneurship (State Council of the People's Republic of China, 2021).

3.3 EE in Colleges and Universities

3.3.1 Concepts of EE in China

Influenced by global EE trends and taking into consideration China's current situation, the development of innovation and EE in China takes the approach of "opening to all students, teaching by categories, combining with disciplines and strengthening practice" (MOE, 2012; State Council, 2015).

- **Opening to all students:** Every student should have the opportunity to access EE, regardless of their academic background.
- **Teaching by categories:** HEIs should provide different types of EE and support resources, so as to meet students' different needs.
- **Combining with disciplines:** The development of EE beyond business schools, with different disciplines offering EE.
- **Strengthening practice:** The teaching process should pay more attention to practice by introducing experiential learning and internship, establishing makerspaces, incubators, and science parks, etc.

Therefore, EE in China comprises not only an entrepreneurship course, degree, or an entrepreneurial practice involving a few people, but it is a systematic project integrated into the whole process of talent cultivation (Mei & Symaco, 2022). Allowing all students access to EE is in line with the needs of solving complex problems and enhancing social and systematic skills against the backdrop of the Fourth Industrial Revolution. However, in addition to cultivating students' professional skills, HEIs should highlight the cultivation of high-order skills such as critical thinking, creativity, initiative, leadership, social influence, entrepreneurial ability, and cultural agility (WEF, 2018; Aoun, 2017).

3.3.2 Curriculum Development of EE

Traditional EE courses were mostly provided by business schools, whose design was mainly based on three approaches: the core elements constituting a business plan; the steps in the entrepreneurial process; and the life cycle of a venture (Morris et al., 2013, pp. 59–60).With the rising importance of innovation and entrepreneurship, EE courses have moved beyond business schools, with interdisciplinary courses such as "Entrepreneurship and the Arts", "Media Entrepreneurship, Entrepreneurship for Psychologists", etc. developed and accessible to students across campus who are interested in entrepreneurship (Morris et al., 2013, pp. 73–74). Based on Johannisson's (1991) entrepreneurial learning theory, Souitaris et al. (2007) conceptualized that entrepreneurial learning comprises five elements: know-why (values and motivation of entrepreneurs), know-what (knowledge of what needs to be done), know-how (practical abilities and skills), know-who (awareness of social networks and the ability to use them), and know-when (experience and intuition about when to take action).

According to the MOE's statistics, around 28,000 entrepreneurship courses have been developed nationally, and nearly 6.3 million students have accessed entrepreneurship courses (Liu, 2019). The MOE provides financial support for university teachers to set up free online entrepreneurship courses for students. As of August 1, 2021, there are 433 online entrepreneurship courses available on the

Table 3.2 Intensive training programme for innovation and entrepreneurship at Zhejiang University (3rd version)

Time	Courses	Credits
Spring and summer semesters of second year (nine credits)	Management	3
	Economics	2
	Entrepreneurial strategic management	2
	Team communication and leadership	2
Autumn and winter semesters of third year (seven credits)	Market survey and analysis	1
	Entrepreneurship marketing	2
	Business model	1
	New product development and project management	2
	Entrepreneurship plan design (1)	1
Spring and summer semesters of third year (six credits)	Startup law and intellectual property	2
	Entrepreneurial valuation and financing	2
	Entrepreneurial finance	1
	Entrepreneurship plan design (2)	1
Autumn and winter semesters of fourth year (four credits)	Entrepreneurship lectures	2
	Entrepreneurship plan design (3)	2

Source ZJU ITP (2019)

massive open online course platform, iCourse, 52 of which have been recognized by the MOE as top-quality courses.

For example, Zhejiang University (ZJU) has introduced a pyramid-style curriculum system. The bottom of the pyramid represents entrepreneurship modules open to all students, including introductory courses on entrepreneurship. The middle of the pyramid are courses that combine specialized knowledge with entrepreneurship elements. At the top of the pyramid is the ZJU minor program for entrepreneurship—Intensive Training Programme for Innovation and Entrepreneurship (ITP). Launched in 1999, the ITP is open to sophomores from different schools, except from the School of Management, with an annual enrollment of 40 students (Mei & Symaco, 2022). The ITP courses are detailed in Table 3.2.

3.3.3 *Business Plan Competitions*

In mainland China, the first Business Plan Competition was organised by Tsinghua University in 1998, which also marked China's first attempt at EE. The first "Challenge Cup" National Business Plan Competition was held in the following year, also by Tsinghua University, which saw participants from more than 100 universities and colleges.

Table 3.3 Internet+ innovation and entrepreneurship plan competition

Year	Organizers	No. of participating HEIs	No. of teams (thousand)	No. of participants (thousand)
2015	Jilin University	1,878	57	200
2016	Huazhong University of Science and Technology	2,110	120	550
2017	Xidian University	2,241	370	1,500
2018	Xiamen University	2,278	640	2,560
2019	Zhejiang University	4,093	1,090	4,570
2020	South China University of Technology	4,186	1,470	6,310

Source Lu (2019)

Nowadays, the most influential entrepreneurship plan competition in China is the College Students' Internet + Innovation and Entrepreneurship Competition (Internet+).In 2015, the MOE called for a new entrepreneurship plan competition that could combine the new generation of information technologies, such as mobile internet, cloud computing, big data, the internet of things, with the industry. The aim was to cultivate new products, services, formats, and models for the internet, as well as to promote public service innovation by deeply integrating the internet with education, medical treatment, and the community (MOE, 2015). Since then, Internet + has attracted the most talented college students and has dramatically improved the innovative and entrepreneurial culture in HEIs. The competition lasts six months, typically from June to November each year (Table 3.3).

The competition serves three purposes: (1) Promoting entrepreneurship through competition and cultivating a new force for innovation and entrepreneurship; (2) Promoting EE through competition and exploring new ways of quality education; (3) Promoting entrepreneurship activities through competition and building a new platform for technology transfer (MOE, 2020).

3.3.4 Organizational Structure of EE

The organizational structure addresses issues like where the entrepreneurship program should be housed, how it will be led, to whom it will report, how it is staffed, and how it operates from a budgetary standpoint (Morris et al., 2013, pp. 20–21).

3.3.4.1 Strategy and Leadership

After the issue of the State Council (2015) circular, EE became a university-wide strategy in many HEIs in China. As requested by the State Council (2015), HEIs should assume the primary responsibility of innovation and entrepreneurship education (IEE), integrate IEE into the institutional reform agenda, and establish a leading team for IEE with a president as the leader, deputy president as vice leader, and heads of relevant departments as members.

3.3.4.2 Development of Colleges of Entrepreneurship

Driven by the central government, colleges of entrepreneurship have become the dominant structure for university-wide EE in China's HEIs since 2015. More than 70 percent of the first 99 "model universities of EE" awarded by the MOE have established or intend to establish colleges of entrepreneurship (Mei & Symaco, 2022; Zhu, 2017). In Zhejiang Province alone, 102 HEIs have since established such colleges (Mei & Symaco, 2022; Wu, 2017). The following section will examine the colleges of entrepreneurship in Zhejiang.

Currently, there are three ways in which Zhejiang's colleges of entrepreneurship are organized (Mei & Symaco, 2022). First, as an independent (standalone) college such as that of Wenzhou University; and second, as part of a university's pre-existing honors college. For instance, prior to the government's mandate to create EE programs, Zhejiang Gongshang University's (ZGU) Honors College already existed, providing high-quality innovation programs for undergraduate students. Following the mandate, ZGU's Honors College created a bi-program, one of which was its innovation programs and the second is the College of Entrepreneurship, which provides university-wide EE. The Zhejiang Agriculture and Forestry University follows this second model.

The third model is a "virtual" college handled by the university's Students Department administration, which is then coordinated by relevant colleges/departments within the university. The latter can be any of the colleges/departments within the university. For example, the College of Education can be coordinating an EE program in cooperation with the university's Student Department. Universities following this third model would include, among others, Zhejiang University, Hangzhou Normal University, Ningbo University, and the Chinese Academy of Arts. The purpose of this model is to enhance the visibility of EE at the university level and make university-wide EE more institutionalized, while the broader scope includes overall planning, resource allocation, and implementation assessment, among others.

3.3.4.3 Improvement of Institutional Guarantee

Based on the policies of the State Council and the MOE, HEIs have also implemented plans for IEE reform. Policies such as a flexible school system for students who start their own business, a credit accumulation and recognition system, a technology transfer system, etc., are introduced to encourage college students to pursue an entrepreneurial career. For example, Shanghai Jiaotong University allows undergraduates and graduates to suspend schooling for two years and one year respectively if they start their own business. Peking University offers information on IEE in its *Annual Report of Teaching Quality and Annual Report on Graduates' Employment.* It regards the quality of IEE as one of the most important indicators for evaluating the quality of teaching and to assess the leaders of colleges and departments (Peking University, 2016).

3.3.5 EE Research

In the past two decades, the research of IEE can be regarded as one of the fastest fields in China. In order to understand the themes and trends of IEE research in China, scholars adopt bibliometric analyses. For example, Huang et al. (2016) conducted a bibliometric review of 894 papers from 37 pedagogic CSSCI journals in China between 1997 and 2016. The study found five research themes: EE and teaching; comparative study of EE; college students' entrepreneurial intention; college students' entrepreneurial quality and ability; and entrepreneurship environment. Xu (2019) produced comprehensive statistics of EE research outcomes in China from 2009 to 2018. According to his research, in that period, Chinese scholars published 3,405 original articles in CSSCI journals (based on a search of China's Journal Network Publishing Database) and 305 original articles in SCI/SSCI journals (based on a search of Web of Science Core Collection Database). Furthermore, national financial support for EE research can indicate the importance of the field to a country. Table 3.4 shows that research projects supported by the China National Social Science Fund (CNSSF) and China National Natural Science Fund (CNNSF) have increased in the past decades.

Through keywords co-occurrence, Xu (2019) pointed out six hot research themes between 2009 and 2018: college students' entrepreneurial intention; higher education reform and IEE; entrepreneurial university and IEE; entrepreneurship policies; relationship between employment and entrepreneurship; and the role of makerspaces in college students' entrepreneurship. Based on this, Xu also proposed that in the future more studies should focus on conceptual and institutional guarantee; the establishment of EE; the integration of EE and professional education; social entrepreneurship and EE; the evaluation and effectiveness of EE; and the convergence of entrepreneurial psychology and EE (Xu, 2019).

Table 3.4 Statistics of EE research in China (2009–2018)

Types		2009	2010	2011	2012	2013	2014	2015	2016	2017	2018	Total
Articles	CSSCI	188	244	280	261	244	248	407	480	578	475	3,405
	SCI/SSCI	8	13	14	15	19	22	30	41	66	77	305
Research projects	CNSSF	12	18	16	17	18	13	13	24	24	22	177
	CNNSF	5	12	14	17	11	17	20	23	19	20	158
	MOE projects	4	10	2	0	5	4	15	8	22	8	78
	Total	21	40	32	34	34	34	48	55	65	50	413

Source Xu (2019)

3.4 Support System of EE

The supportive environment and other factors (educator, curriculum, etc.) of EE are complementary. Building a support system as an effective and sustainable approach to contributing to the entire ecosystem is crucial for university EE. This also has an undeniable impact on graduates in terms of their career choices and entrepreneurial behaviors. The EE support system can be analyzed in terms of stakeholders, resources, infrastructure, and culture (Fetters et al., 2010).

3.4.1 Stakeholders

Stakeholders refer to all groups that *can affect or be affected* by certain activities. In the context of IEE, they are directly or indirectly influencing IEE either through active involvement in the provision of education, or as recipients of education (Bischoff & Volkmann, 2018; Freeman, 2010, p. 9). Therefore, internal stakeholders of IEE comprise university leaders, managers, faculty members, students, etc., while external stakeholders include the government, enterprises, investors, incubators, etc.

3.4.2 Resources

3.4.2.1 Resources for Encouraging Creative Ideas

In 2007, the MOE launched the "National Innovation and Entrepreneurship Training Program for College Students" (National Innovation Program). In accordance with the principles of "interest-driven, independent practice and focusing on the process", the National Innovation Program encourages college students to carry out innovative practice and encourages HEIs to conduct comprehensive reforms in terms of teaching content, curriculum system, and practical procedures, so as to improve the innovative and entrepreneurial ability of college students. It is estimated that more than 900,000 college students from around 1,000 HEIs have participated in the program, and a total of 220,000 national projects have been funded, covering all disciplines, with the funding support of about RMB3.7 billion (MOE, 2019). The National Innovation Program has become a basic project for all college students.

3.4.2.2 Resources for Graduate Entrepreneurship

The Shanghai Technology Entrepreneurship Foundation for Graduates (EFG) was established in 2006. As the country's first nonprofit public foundation that motivates college students to undertake technology entrepreneurship, it is committed to spreading entrepreneurship culture and supporting entrepreneurial practice. The EFG has set up 23 sub-foundations in conjunction with various colleges and universities, districts, counties, and incubators. Being charity-based and market-oriented are the two engines for mechanism innovation. Since its inception, it has been carrying out pivotal activities in fields such as entrepreneurial culture, entrepreneurial studies, EE, and project incubation.

The EFG provides comprehensive resource support for students and graduates (within five years of graduating), including advocacy, education, funding, and services. Figure 3.4 shows that the core of the resource support system is the Angel Fund, which consists of two modes: the "Entrepreneurship Eaglet Program", which provides interest- and mortgage-free startup loans for up to RMB500,000; and the "Entrepreneurship Eagle Program", which offers equity investment of up to RMB800,000, without the need to share bonuses and to withdraw at the original price. It is estimated that by the end of April 2021, these two modes of Angel Fund accepted a total of 11,430 applications for entrepreneurship projects, provided funds for 3,274 projects, and created more than 30,000 new jobs (STEFG, 2021).

In 2012, the China Higher Education Student Information and Career Center and the Huatu Education Group jointly invested RMB10 million to establish the Huatu Entrepreneurship Fund for University Students, which would provide more than RMB2 million for student startups as a nonprofit support. Through this fund, students who want to start a business are provided with financial support and professional guidance. By 2014, 3,177 student entrepreneurship projects from 1,178 universities in 31 cities applied for the Huatu fund. The projects range from e-business, creative culture, and education services, to energy conservation, environmental protection, and software technology. Companies also provide financial support for university EE through various ways, like electing venture awards and sponsoring competitions.

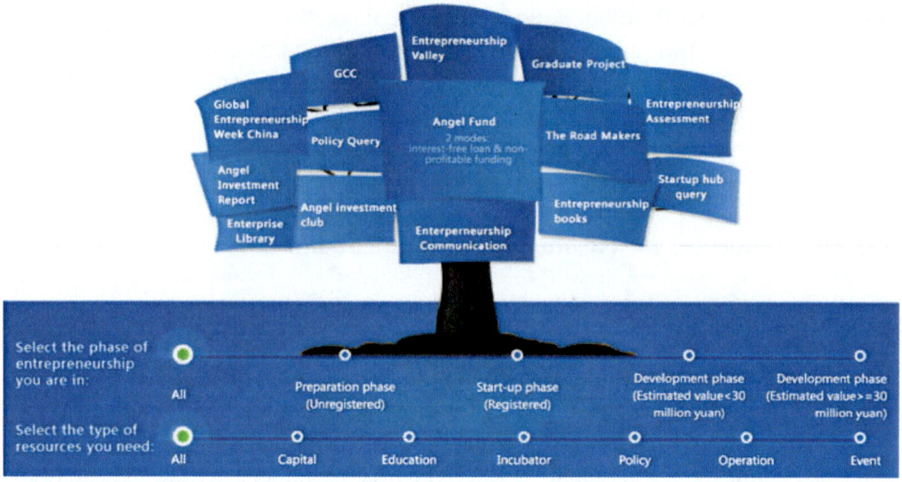

Fig. 3.4 Entrepreneurship resources provided by the Shanghai Technology Entrepreneurship Foundation for Graduates

3.4.3 *Infrastructure*

3.4.3.1 Development of National Incubators and University Science Parks

Technology business incubators are new economic organizations aimed at providing infrastructure and a series of supportive services for startups. The main purpose of incubators is to help startups grow, reduce risks and costs, and finally achieving independent development.

Incubators have been around for many years and they are referred to by other terms such as innovation centers, technology/science parks, etc. The Donghu Entrepreneurship Center was the first incubator in China, established in 1987. Over the past three decades, incubators have entered a stage of stable development. According to China Venture Research (Ministry of Science and Technology of the People's Republic of China, 2020a, b), at the end of 2019, there were 5,206 technology business incubators (TBIs) in China, 1,177 of which were national incubators (see Fig. 3.5 and Table 3.5).

University science parks are also an important infrastructure for supporting IEE in HEIs. In China, university science parks went through three stages: initial exploration (1989–1998), normative development (1999–2009), and steady development (since 2010). At present, China has a three-level university science park system run by national, provincial, and institutional HEIs respectively. At the end of 2018, there were 115 national university science parks in China, with a total site area of 7.086 million square meters (MOST, 2019). A total of 10,127 enterprises are incubated in the national science parks, with an employment of 128,000 and an income of RMB32.5 billion (MOST, 2019).

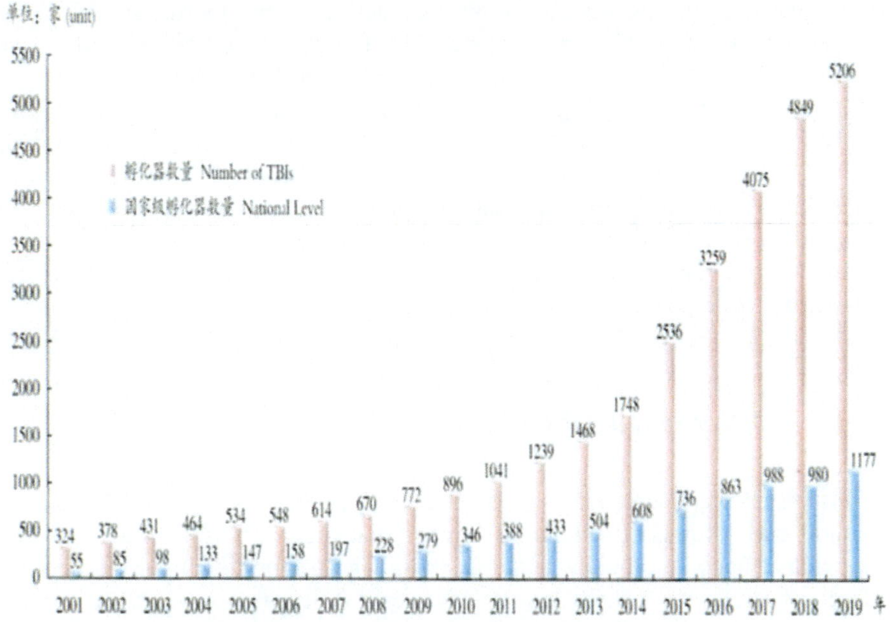

Fig. 3.5 The number of national incubators in China

Table 3.5 General statistics of funding status of TBI tenants

	2018	2019	Growth Rate
R&D investment of tenants	726.7	705.0	−3.0
Accumulated VC investment to Tenants	2755.9	2606.0	−5.4
Accumulated financial subsidy to tenants	220.1	238.1	8.2

Source Ministry of Science and Technology of the People's Republic of China (MOST) (2020a, b)

3.4.3.2 Tsinghua x-lab

Tsinghua's x-lab is a university-based education platform designed to foster student creativity, innovation, and entrepreneurship. It brings together Tsinghua students, faculty, alumni, experienced entrepreneurs, investors, and experts from across society.

'X' refers to the mission of pursuing the unknown at the intersection of multiple academic disciplines and needs. 'Lab' emphasizes the experiential, collaborative, and action-based process that is core to x-lab's approach to learning. Officially launched in April 2013, x-lab operates under the auspices of the School of Economics and Management. It links 14 schools and departments across Tsinghua University, including Mechanical Engineering, Natural Sciences, Information Science and Technology, Arts and Design, Medicine, Aerospace, Environment, Architecture, Materials

Science and Engineering, Public Policy and Management, Journalism and Communication, Law, and Engineering Physics. The Tsinghua Entrepreneur and Executive Club and Tsinghua Science Park are strategic founding partners.

Since its establishment, x-lab has organized lectures, competitions, exchanges, and other entrepreneurial activities for students, attracting more than 30,000 students from Tsinghua University and beyond (Tsinghua University, 2021). In addition, more than 1,500 creative, innovative, and entrepreneurial projects by Tsinghua University students and alumni have joined x-lab. Through the cultivation of x-lab, the total financial value of registered companies has exceeded RMB3 billion (Tsinghua University, 2021).

3.4.4 Entrepreneurial Culture

Confucianism is deep-rooted in Chinese culture. The saying, "Officialdom is the natural outlet for good scholars", implies that high-performing students will be selected to become officials, and only those who perform badly academically will choose business to make a living. However, in the past two decades, due to the implementation of IEE and the improvement of entrepreneurship environment, entrepreneurial culture in China has improved dramatically.

Figure 3.6 shows that, in the past two decades, government policies (4.79, 17/54), internal market dynamic (6.67, 5/54), and physical infrastructure (7.4, 6/54) have performed well, while EE in schools (5.27, 15/54), R&D transfer (4.04, 28/54), entrepreneurial finance (4.6, 27/54), and commercial and legal infrastructure (4.23, 47/54) still have room for improvement (GEM, 2019, p. 73).

3.5 Characteristics of EE

There is no one-size-fits-all model for IEE in different HEIs in China. In order to promote the efficacy of policies, China's IEE gradually developed its own characteristics.

3.5.1 Implementing Policy Experimentation of IEE

Due to the regional and institutional differences, China has adopted policy experimentation as an important means of introducing and testing innovative policy options (Han, 2020). The "experimental governance" enables state–university interactions and power negotiations that create and maintain strategic space for consensus-building (Han, 2020; Ning, 2014).

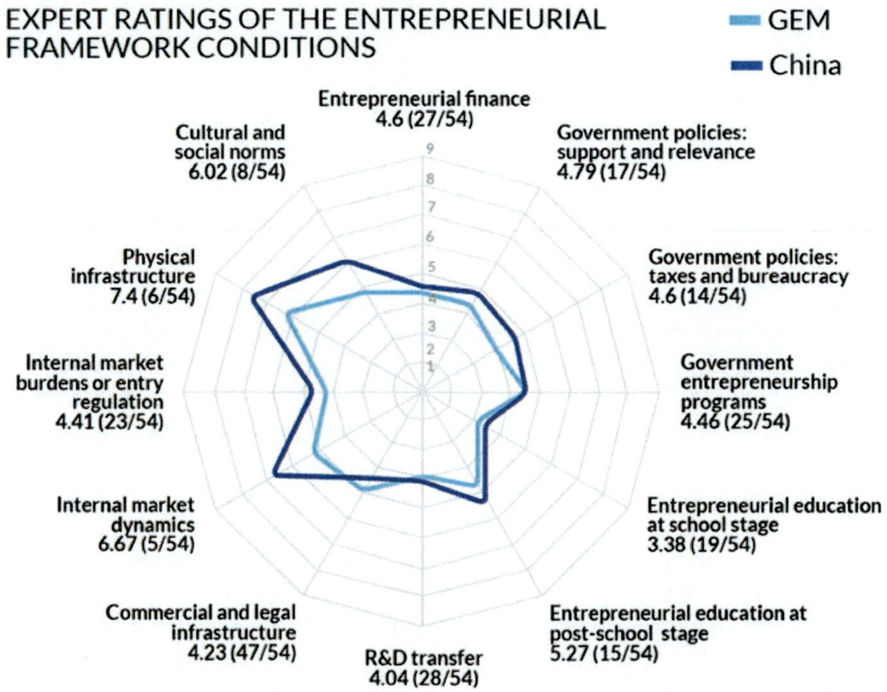

Fig. 3.6 Expert rating of the entrepreneurial framework conditions of GEM. *Source* GEM (2019, p. 73). EFCs scale: 1 = very inadequate insufficient status, 9 = very adequate sufficient status

China has adopted the same logic to promote IEE. Based on several years of institutional exploration, nine universities were selected by the MOE to participate in the National Entrepreneurship Education Pilot Program in as early as 2002. The nine universities include Tsinghua University, Beijing University of Aeronautics and Astronautics, Renmin University of China, Heilongjiang University, Shanghai Jiaotong University, Nanjing University of Finance and Economics, Wuhan University, Xi'an Jiaotong University, and Northwest Polytechnic University. With pilot support from the MOE, they can reconstruct a cross-campus EE system and explore new approaches. The nine universities adopt different models for EE: classroom-based (Renmin University of China), practice-oriented (Beijing University of Aeronautics and Astronautics), and hybrid (Shanghai Jiaotong University) (Zhou & Xu, 2012).

In 2008, the MOE established 100 experimental zones for cultivating innovative talents, 32 of which are directly related to EE. Since 2016, the MOE has recognized 200 HEIs with "Typical Experience of Entrepreneurship Education", setting an example for other HEIs (MOE, 2019). Though the MOE retains the ultimate authority for legitimatizing, selecting, and expanding policy experiments (Han, 2020), the awarding of HEIs with "Typical Experience of Entrepreneurship Education" has shown positive impact for identifying best practices and promoting experience exchange among different HEIs. Through policy preference, extra financial

Table 3.6 HEIs recognized as Typical Experience of EE

	2016	2017	2018	2019
No. of HEIs under the administration of central departments	19	13	9	8
No. of provincial four-year HEIs	25	30	33	31
No. of Higher Vocational Colleges	6	7	8	11
Total	50	50	50	50

Source MOE (2019)

support, and publicity, these HEIs have become the "vanguard" of EE in China (Mei, 2020, pp. 145–146) (Table 3.6).

In the area of EE, the central government distinguishes general HEIs from pilot HEIs or HEIs with "Typical Experience." It authorizes the latter to carry out policy experiments and provides them with extra financial support or policy preference to encourage pilot HEIs to explore innovative approaches (Yang, 2013). In the future, general HEIs or latecomers to IEE in China need to: (1) move from being government-driven to promote their internal driving forces to conduct IEE; (2) while learning from the experiences of model colleges and universities, general HEIs should conduct IEE based on their regional and institutional characteristics; (3) encourage entrepreneurs, investors, and alumni to actively participate in IEE, and reform processes so as to promote participation.

3.5.2 Conducting University-Wide IEE

The development of EE adopts a university-wide logic—i.e., open to all students and mobilizing different disciplines to integrate EE into their curriculum. On the one hand, university-wide IEE reflects international trends. For example, in 2003 and 2006, the Kauffman Foundation in the US implemented two rounds of the Kauffman Campus Initiative, cumulatively providing USD45 million and another USD148 million of matching funds to promote cross-campus EE (Morris et al., 2013; Schneider, 2015, p. 15). The principles for selecting HEIs are: (1) whether a school can make entrepreneurship a common and accessible activity for all students; (2) the level of involvement of the president or chancellor; (3) the ability to generate matching funds; (4) whether the school can serve as a model for other colleges and universities; (5) the relative strength of the innovative approaches; and (6) the likelihood that the initiative would change campus culture and produce a sustainable entrepreneurial spirit on campus (Morris et al., 2013, p. 246). The university-wide concept tries to make IEE a unique experience for students from different academic backgrounds and encourages interdisciplinary development.

On the other hand, China's IEE has its own characteristics. University-wide IEE in China not only cultivates innovative and entrepreneurial talents, but has also been regarded as a breakthrough for reforming the entire higher education system (Mei,

2020). There have been efforts to change the current framework, but there is strong inertia from institutions (Battilana et al., 2009; DiMaggio & Powell, 1983).

In this context, many institutional entrepreneurs have played key roles to explore unique approaches to promote university-wide EE in China's context. The international experience shows that there are two main structural models of university-wide EE: the magnet model and the radiant model (Streeter et al., 2011). The former emphasizes that one college takes all the responsibility of IEE in HEIs, while the latter encourages different colleges to be involved in IEE. In China's colleges and universities, actors are both constrained and enabled by institutional frameworks. However, they are capable of using these frameworks to pursue their own interests as well as to challenge and change frameworks if necessary (Kirst & Stevens, 2015, pp. 28–29). They possess the capacity to exercise *agency*—the ability to act independently, resist and even change the relational contexts within which they are embedded (Kirst & Stevens, 2015, pp. 28–29; Lawrence & Suddaby, 2006). Based on the spontaneous exploration of a few colleges and universities at the beginning, institutional entrepreneurs in China's colleges and universities gradually developed the College of Entrepreneurship Model to promote university-wide EE. As an achievement of institutional entrepreneurship, the College of Entrepreneurship Model in China is neither a magnet model nor a radiant model (Streeter et al., 2011), but a combination of the two. Not only can it overcome China's institutional obstacles of implementing university-wide IEE in a short time, but also take the advantage of the two models and act as a hub for university-wide efforts to promote EE (Mei & Symaco, 2022). It can help negotiate with different stakeholders to more efficiently participate in IEE in China.

3.5.3 Main Challenges of IEE in China

The past 20 years have witnessed the boom of IEE programs in China's HEIs, and the support system of IEE has also improved dramatically. Due to the expansion of the higher education system and the fast development of IEE since the end of the 1990s, the proportion of under-educated entrepreneurs in China has gradually decreased. At the same time, both the proportion of highly educated entrepreneurs and the number of high-income individuals starting businesses have increased (GEM, 2019, p. 73).However, the Global Entrepreneurship Monitor (GEM) (2019) indicated that recognition of entrepreneurial abilities decreased while the fear of failure gradually increased. This is consistent with statistics from the Chinese College Graduates' Employment Annual Report (Mycos Institute, 2019), which stated that the proportion of 2018 graduates starting a business within half a year from graduation was 2.7%, the survival rate of startups within three years after graduation was 44.8%, and the overall success rate of undergraduates' startup was less than 5 percent (Lin, 2019).

Besides, in the past two decades, the fast development of IEE is mainly due to government-driven efforts, and multiple-stakeholder involvement should be further stimulated. According to the 2015 *Report on Chinese Youth Entrepreneurship*, 64.2%

of the respondents mentioned the lack of financial support as the main difficulty in entrepreneurship (MOHRSS, 2016). For 2017 university graduates, 51 percent of the funds for starting their own business came from parents/relatives and friends, 25 percent from personal savings, 4 percent from government funds or preferential loans, and 7 percent from bank loans (Mycos Institute, 2018). These statistics show that, on the one hand, Chinese college students lack the knowledge and skills in getting financial support and have a weak awareness of actively seeking external funds, which result in poor and ineffective connection between external funds and college students' entrepreneurship activities. On the other hand, support for college students' entrepreneurship projects is constrained by complicated conditions, procedures, and costs, which impedes college students to apply for the funds. Therefore, while providing diversified financial support for college students, it is of utmost importance to reduce the financial threshold, simplify procedures, and improve the whole system, so as to help college students turn ideas into action.

3.6 Toward a Higher Quality of EE

How IEE in HEIs be further promoted in China? How can bottom-up initiatives be stimulated to develop IEE? How can IEE better satisfy the requirements of digital development and sustainable development as well as the 14th Five-Year Plan and Vision 2035 for National Economic and Social Development of the People's Republic of China? These questions should be considered systematically.

3.6.1 Enhancing Policy Effectiveness

Little research has been done to answer questions like what policy areas have the greatest impact on EE and graduate entrepreneurship. By studying entrepreneurs from 22 countries, Monitor Group (2009) indicates that some recommendations for the promotion of entrepreneurship are far less effective than claimed. The study further suggests that policies should focus on incentives, not assistance, and that they should pay more attention to cultivating entrepreneurial mindsets and teaching entrepreneurial skills. In the past decade, the Chinese government has issued many policies to promote EE and improve the entrepreneurial atmosphere. However, most of these policies are quite general and lack assessment for effectiveness. More research is needed to evaluate the implementation and effectiveness of IEE policies in the near future. The government should also issue more specific policies like encouraging professional teachers to attend IEE courses; increasing IEE funding from different channels; and encouraging Higher Vocational Colleges to conduct diversified IEE programs.

3.6.2 Promoting Diversified IEE Programs

Since the late 1990s, the development of IEE within HEIs has adopted the bottom-up model. Some entrepreneurial champions or faculty members created small programs to explore the cultivation of entrepreneurial talents, without additional budget. In the last five years, driven by the intense IEE policies, IEE programs in HEIs have adopted a top-down style, which to a great extent leads to a homogenous development of IEE. In the future, HEIs should promote diversified and differential development by formulating IEE programs with strong disciplines and regional characteristics.

Technology EE. The national strategic plan and industrial policy *Made in China 2025 (MIC 2025)* puts forward 10 key areas of the manufacturing industry: information technology; robotics; green energy and green vehicles; aerospace equipment; ocean engineering and high-tech ships; railway equipment; power equipment; new materials; medicine and medical devices; and agriculture machinery. All these place higher requirements for the transformation of IEE in colleges and universities. It is the responsibility of colleges and universities to cultivate more innovative and entrepreneurial talents, so as to use these technologies to solve medical, urban traffic, and manufacturing problems.

Digital EE. With the development of digital technologies, digital entrepreneurship is attracting world attention (Ngoasong, 2018). Digital entrepreneurship refers to "the reconciliation of traditional entrepreneurship with the new way of creating and doing business in the digital era" (Le Dinh et al., 2018), which should be recognized as "augmented entrepreneurship" rather than a subcategory of entrepreneurship (Sahut et al., 2019). In China, the digital economy includes digital industrialization and the digitalization of traditional industry. The scale of China's digital economy reached RMB39.2 trillion in 2020, accounting for 38.6 percent of its GDP and representing a nominal year-on-year growth of 9.7% (CAICT, 2021). In July 2020, 13 central departments, including the National Development and Reform Commission, issued the *Opinion on Supporting the Healthy Development of New Formats and New Models, Activating the Consumer Market and Promoting the Expansion of Employment*. This proposed supporting the development of 15 new formats and new models of economy, including online education, online medical care, digital governance, industrial platform development, digital transformation of traditional enterprises, "virtual" industrial parks and industrial clusters, and the "unmanned" economy (NDRC, 2020). The cultivation of innovative and entrepreneurial talents should make full use of digital technology, not only encouraging students to establish digital enterprises, but also cultivating students with entrepreneurial digital mindsets and competencies.

Social EE. IEE can produce not only economic value, but also social value. In a globalized world, humans are facing common global social, economic, and environmental challenges. The 17 Sustainable Development Goals (SDGs) adopted by the United Nations member states in 2015 provide a blueprint for further development, and they attach great importance to innovation and entrepreneurship. For example, SDG 4.4 proposed that "By 2030, substantially increase the number of youth and adults who have relevant skills, including technical and vocational skills, for employment, decent jobs and entrepreneurship". SDG 8.3 calls for "Promoting development-oriented policies that support productive activities, decent job creation, entrepreneurship, creativity and innovation, and encourage formalization and growth of micro-, small- and medium-sized enterprises including through access to financial services" (United Nations, 2015). IEE is also an important platform to cultivate students' social responsibility and ability to use entrepreneurial principles to solve domestic social challenges. For example, the "Rural Revitalization Strategy" and the "Targeted Poverty Alleviation Strategy" are two key strategies to promote rural development and address issues of poverty in China. Colleges and universities should guide students to use their professional skills and entrepreneurial ability to be involved in solving these challenges. Some provinces in China have attached importance to social EE. For example, since 2018, Zhejiang Province has been conducting the "College Students' Creative Competition for Rural Revitalization", which encourages college students to form business plans targeting issues in rural regions including agriculture, cultural industry, tourism, and social services (Lu, 2019).

3.6.3 Promoting Stakeholder Involvement

HEIs should rely on the IEE faculty from business schools, cross-disciplinary faculty from collaborative technology innovation and design-thinking disciplines, to establish a teacher organization mechanism of cross-border learning. This is to promote the creation of innovation and entrepreneurship undergraduate courses, certificates, majors, and degrees.

It is important to employ entrepreneurs and investors to serve as entrepreneurial mentors, establishing cooperation with investment institutions and entrepreneurial professional service institutions, so as to provide a variety of support for entrepreneurial teams. Besides, with increasing global cooperation in IEE, it is necessary to explore a diversified and international joint training model for innovative and entrepreneurial talents; to jointly organize international creation, innovation, and entrepreneurship competitions; and to launch international entrepreneurship internships.

References

Antal, N., Kingma, B., Moore, D., & Streeter, D. (2014). University-wide entrepreneurship education. In S. Hoskinson & D. F. Kuratko (Eds.), *Innovative pathways for university entrepreneurship in the 21st century* (pp. 227–254). Emerald Group Publishing Limited.

Aoun, J. E. (2017). *Robert-proof: Higher education in the age of artificial intelligence*. The MIT Press.

Battilana, J., Leca, B., & Boxenbaum, E. (2009). How actors change institutions: Towards a theory of institutional entrepreneurship. *The Academy of Management Annal, 3*(1), 65–107.

Bischoff, K., & Volkmann, C. K. (2018). Stakeholder collaboration in entrepreneurship education: An analysis of the entrepreneurship ecosystems of European higher education institutions. *Journal of Technology Transfer, 43*(1), 20–46.

China Academy of Information and Communication Technology. (2021, Apirl). *White paper on China's digital economy development*. Author. [In Chinese]

China Labor Security Newspaper. (2018, April 19). *The earlier the better the professional skills development bill to be made*. Retrieved from http://www.mohrss.gov.cn/zynljss/gzjl/201804/t20180419_292596.html [In Chinese].

China's post-90s entrepreneurs. (n.d.). *China Daily*. Retrieved from http://www.chinadaily.com.cn/china/entrepreneursinchina/

CPC Central Committee. (2020, November 3). *Outline of the 14th Five-Year Plan and Vision 2035 for National Economic and Social Development of the People's Republic of China*. Retrieved from http://www.gov.cn/zhengce/2020-11/03/content_5556991.htm [In Chinese].

DiMaggio, P. J., & Powell, W. W. (1983). The iron cage revisited: Institutional isomorphism and collective rationality in organizational fields. *American Sociological Review, 48*(2), 147–160.

European Commission. (2018). *Council recommendation of 22 May 2018 on key competences for lifelong learning*. Author. Retrieved from https://eur-lex.europa.eu/legal-content/EN/TXT/PDF/?uri=CELEX:32018H0604(01)&from=EN

Feng, Y. F. (2013). *Is it really that difficult for college graduates to get a job? Evidence based on employment rate and income* (Master's thesis, Zhejiang University, Hangzhou, China). Retrieved from https://kns.cnki.net/KCMS/detail/detail.aspx?dbname=CMFD201401&filename=1013339953.nh [In Chinese].

Fetters, M. L., Greene, P. G., Rice, M. P., & Butler, J. S. (2010). *The development of university-based entrepreneurship ecosystems: Global practices*. Edward Elgar Publishing.

Freeman, R. E. (2010). *Strategic management: A stakeholder approach*. Cambridge University Press.

Global Entrepreneurship Monitor. (2019). *Global entrepreneurship monitor global report 2018/19*. Retrieved from https://www.gemconsortium.org/file/open?fileId=50213

Han, S. (2020). Policy experimentation and power negotiation in China's higher education reforms. *Higher Education, 79*(2), 243–257.

Huang, Z. X., Li, Y. Y., & Liu, M. Y. (2016). Research on twenty-years' entrepreneurship education in China: Research focuses, trends and evolution paths—A bibliometric review of 37 pedagogic CSSCI journals. *Educational Research, 39*(1), 64–73 [In Chinese].

International Labor Organization. (2015). *World employment social outlook trends 2015*. Author. Retrieved from https://www.ilo.org/wcmsp5/groups/public/---dgreports/---dcomm/---publ/documents/publication/wcms_337069.pdf

Johannisson, B. (1991). University training for entrepreneurship: Swedish approaches. *Entrepreneurship & Regional Development, 3*(1), 67–82.

Kirst, M. W., & Stevens, M. L. (2015). *Remaking college: The changing ecology of higher education*. Stanford University Press.

Know About Business. (2011, September 15). *Introduction to know about business China*. Retrieved from http://chuangye.cyol.com/content/2011-09/15/content_4894576.htm [In Chinese].

Lawrence, T. B., & Suddaby, R. (2006). Institutions and institutional work. In S. R. Clegg, C. Hardy, T. B. Lawrence, & W. R. Nord (Eds.), *The Sage handbook of organization studies* (2nd ed., pp. 215–254). Sage.

Le Dinh, T., Vu, M. C., & Ayayi, A. (2018). Towards a living lab for promoting the digital entrepreneurship process. *International Journal of Entrepreneurship, 22*(1), 1–17.

Lin, C. H. (2019, October 29). Reconstruction of innovation and entrepreneurship education concept. *Guangming Daily* (p. 14) [In Chinese].

Lindner, J. (2020). *Entrepreneurial learning for TVET institutions: A practical guide*. UNESCO Publishing.

Lu, J. (2019, November 1). *Zhejiang college students' creative competition for rural revitalization has achieved fruitful results.* Retrieved from https://difang.gmw.cn/zj/2019-11/01/content_33286624.htm [In Chinese].

Lundström, A., & Stevenson, L. A. (2005). *Entrepreneurship policy: Theory and practice.* Kluwer Academic Publishers.

Mei, W. H. (2020). *Innovation of organizational models and operational mechanism of entrepreneurship education in higher education institutions.* Zhejiang University Press [In Chinese].

Mei, W. H., & Symaco, L. (2022). University-wide entrepreneurship education in China's higher education institutions: Issues and challenges. *Studies in Higher Education, 47*(1), 177–193.

Ministry of Education of the People's Republic of China. (2010, May 5). *Opinions on boosting the innovation and entrepreneurship education in higher education institutions and college students' entrepreneurship.* Retrieved from http://www.moe.gov.cn/publicfiles/business/htmlfiles/moe/s5672/201105/xxgk_120174.html [In Chinese].

Ministry of Education of the People's Republic of China. (2012, August 1). *Basic requirements of teaching entrepreneurship education in higher education institutions.* Retrieved from http://www.moe.gov.cn/srcsite/A08/s5672/201208/t20120801_140455.html [In Chinese].

Ministry of Education of the People's Republic of China. (2015, June 2). *Notice on organizing the first China college students' 'internet' innovation and entrepreneurship competition.* Retrieved from http://www.moe.gov.cn/srcsite/A08/s5672/201506/t20150602_189532.html [In Chinese].

Ministry of Education of the People's Republic of China. (2018). *National statistical bulletin on the development of education in 2017.* Retrieved from http://www.moe.gov.cn/jyb_sjzl/sjzl_fztjgb/201807/t20180719_343508.html [In Chinese].

Ministry of Education of the People's Republic of China. (2019a, July 31). *The administrative measures of national innovation and entrepreneurship training program for college students.* Retrieved from http://www.moe.gov.cn/jyb_xwfb/gzdt_gzdt/s5987/2019a07/t2019a0731_393103.html [In Chinese].

Ministry of Education of the People's Republic of China. (2019b, August 5). *Notice on publishing the list of HEIs with typical experience of entrepreneurship education in 2019.* Retrieved from http://www.moe.gov.cn/srcsite/A15/s7063/2019b08/t2019b0822_395517.html [In Chinese].

Ministry of Education of the People's Republic of China. (2020, June 3). *Notice on organizing the sixth China college students' 'internet' innovation and entrepreneurship competition.* Retrieved from http://www.moe.gov.cn/srcsite/A08/s5672/202006/t20200604_462707.html [In Chinese].

Ministry of Human Resources and Social Security of the People's Republic of China. (2016, April 12). *Report on Chinese youth entrepreneurship.* Retrieved from http://www.chinanews.com/cj/2016/04-12/7831083.shtml [In Chinese].

Ministry of Science and Technology of the People's Republic of China. (2019, December 25). *Statistics of university science parks in China in 2018.* Retrieved from https://www.chyxx.com/industry/201912/821872.html [In Chinese].

Ministry of Science and Technology of the People's Republic of China. (2020a, August 27). *Statistical bulletin of national science and technology investment in 2019.* National Bureau of Statistics of China. Retrieved from http://www.stats.gov.cn/tjsj/tjgb/rdpcgb/qgkjjftrtjgb/202107/t20210720_1819716.html [In Chinese].

Ministry of Science and Technology of the People's Republic of China. (2020b, December 7). *2019 Torch statistics manual (technology business incubator and maker spaces).* Retrieved from http://www.chinatorch.gov.cn/fhq/tjnb/2020b12/a4ec6b67553d4c35acc9655bf536ec63.shtml [In Chinese].

Monitor Group. (2009, January). *Paths to prosperity: Promoting entrepreneurship in the 21st century.* Retrieved from https://icma.org/documents/paths-prosperity

Morris, M. H., Kuratko, D. F., & Cornwall, J. R. (2013). *Entrepreneurship programs and the modern university.* Edward Elgar.

Mycos Institute. (2010). *Chinese college graduates' employment annual report (2010).* Social Science Academic Press (China) [In Chinese].

Mycos Institute. (2014). *Chinese college graduates' employment annual report (2014).* Social Science Academic Press (China) [In Chinese].

Mycos Institute. (2018). *Chinese college graduates' employment annual report (2014).* Social Science Academic Press (China) [In Chinese].

Mycos Institute. (2019). *Chinese college graduates' employment annual report (2014).* Social Science Academic Press (China) [In Chinese].

NDRC. (2020). *Opinion on supporting the healthy development of new formats and new models, activating the consumer market and promoting the expansion of employment*. Retrieved from http://www.gov.cn/zhengce/zhengceku/2020-07/15/content_5526964.htm [In Chinese].

Ngoasong, M. Z. (2018). Digital entrepreneurship in a resource-scarce context: A focus on entrepreneurial digital competencies. *Journal of Small Business and Enterprise Development, 25*(3), 483–500.

Ning, S. (2014). Institutional factors in policy experiment from an international comparative perspective. *Expanding Horizons, 2*, 27–33.

Organization for Economic Cooperation and Development. (2018, April). *The future of education and skills: Education 2030.* OECD Publishing. Retrieved from https://www.oecd.org/education/2030/E2030%20Position%20Paper%20(05.04.2018).pdf

Peking University. (2016). *Implementation plan for deepening innovation and entrepreneurship education reform of Peking University.* http://www.moe.gov.cn/s78/A08/gjs_left/s3854/cxcyjy_ssfa/201605/t20160516_244087.html [In Chinese]

Peters, M. A. (2017). The Chinese dream: Xi jinping thought on socialism with Chinese characteristics for a new era. *Educational Philosophy and Theory, 49*(14), 1299–1304.

Qian, Y. Y. (2017, July 5). *Artificial intelligence will wipe out China's educational advantages.* Retrieved from https://www.guancha.cn/TMT/2017_07_05_416682.shtml [In Chinese]

Sahut, J. M., Iandoli, L., & Teulon, F. (2019). The age of digital entrepreneurship. *Small Business Economics 56,* 1159–1169. Retrieved from https://doi.org/10.1007/s11187-019-00260-8

Schneider, M. (2015). *Kauffman Campus Initiative: A study that explores the phenomenon of cross-campus entrepreneurship* (Doctoral dissertation, University of Pennsylvania, Philadelphia, Pennsylvania, the United States). Retrieved from https://www.proquest.com/dissertations-theses/kauffman-campuses-initiative-study-that-explores/docview/1727125381/se-2

STEFG (Shanghai Technology Entrepreneurship Foundation for Graduates). (2021). *About the foundation.* Retrieved from http://www.stefg.org/EN/about/about.aspx.

Souitaris, V., Zerbinati, S., & Al-Laham, A. (2007). Do entrepreneurship programmes raise entrepreneurial intention of science and engineering students? The effect of learning, inspiration and resources. *Journal of Business Venturing, 22*(4), 566–591.

State Council of the People's Republic of China. (2015, May 4). *Implementation opinions on deepening the reform of innovation and entrepreneurship education in higher education institutions.* Retrieved from http://www.gov.cn/zhengce/content/2015-05/13/content_9740.htm?gs_ws=tsina_635673676833950917 [In Chinese]

State Council of the People's Republic of China. (2021, October 12). *Guidance on further supporting college students' innovation and entrepreneurship.* Retrieved from http://www.gov.cn/zhengce/content/2021-10/12/content_5642037.htm.

Streeter, D. H., Kher, R., & Jaquette, J. P., Jr. (2011). University-wide trends in entrepreneurship education and the rankings: A dilemma. *Journal of Entrepreneurship Education, 14*, 75–92.

Tencent Education. (2018, June 11). *Report shows that the proportion of self-employed college graduates has increased significantly in the past decade.* Retrieved from https://new.qq.com/cmsn/20180611/20180611029867.html [In Chinese]

Tongji University. (2021). *Introduction of Tongji Venture Valley.* http://chuangyegu.tongji.edu.cn/index.php?classid=6591

Tsinghua University. (2021). *Introduction of Tsinghua X-Lab.* Retrieved from http://www.x-lab.tsinghua.edu.cn/about.html#xlabjj

Tsinghua University & Fudan University. (2016, October). *Research on the skills gap in China's labor market.* Retrieved from https://www.tsinghua.edu.cn/__local/4/E6/DA/A12EB75B9D564353167D4F107C5_D711D7DB_79EC7D.pdf [In Chinese].

United Nations. (2015). Transforming our world: The 2030 agenda for sustainable development. Retrieved from https://sustainabledevelopment.un.org/content/documents/21252030%20Agenda%20for%20Sustainable%20Development%20web.pdf

Wang, Z. R. (2016). *The history of innovation and entrepreneurship education in China.* Social Sciences Academic Press (China) [In Chinese].

World Economic Forum. (2018, September). *The future of jobs survey 2018.* Retrieved from http://www3.weforum.org/docs/WEF_Future_of_Jobs_2018.pdf

Wu, Q. Q. (2017, October 11). Nearly a hundred colleges and universities set up colleges of entrepreneurship in Zhejiang province. *Zhejiang News.* Retrieved from http://zjnews.zjol.com.cn/zjnews/201710/t20171011_5310611.shtml [In Chinese]

Wu, Y. (2019, December 15). *Innovation and entrepreneurship education: Profound change of education paradigm and new concept of quality.* Retrieved from https://www.sohu.com/a/360697810_120044587 [In Chinese].

Xu, X. Z. (2019). Characteristics and trends of entrepreneurship education researches in China: Based on the visualized analysis of research outcomes from 2009–2018. *China Higher Education Research, 3*, 52–60 [In Chinese].

Yang, H. S. (2013). Dual-track approach to policy experiment: The experience of policy innovation in China. *Chinese Public Administration, 6*, 12–15, 103 [In Chinese].

Zhang, Y. L., & Su, Y. J. (2016). How innovation and entrepreneurship education become a national action policy: Research based on the theory of multiple streams. *Research in Educational Development, 36*(5), 41–47 [In Chinese].

Zhou, M. S., & Xu, H. X. (2012). A review of entrepreneurship education for college students in China. *Administrative Sciences, 2*(1), 82–98.

Zhu, J. D. (2017). Analysis of organizational characteristics of entrepreneurship colleges in universities: Empirical data based on the first group of the innovation and entrepreneurship education reform model universities] *China Higher Education Research, 11*, 49–53 [In Chinese].

ZJU. (2019). *Training program of intensive training program of innovation and entrepreneurship at ZJU (3.0 Version).* Zhejiang University.

Chapter 4
Entrepreneurship Education in Croatia

Slavica Singer

4.1 Context Matters

Context always matters—it is not possible to understand and interpret any event or activity without knowing the circumstances in which they occur. Entrepreneurship education (EE) or entrepreneurship skills started appearing in international policy documents (such as those of UNESCO, OECD, European Union, etc.) in the 1990s, but much earlier in US higher education institutions (HEIs). The first course on entrepreneurship was introduced at the Harvard Business School as early as 1947; in the 1980s, after a lull, entrepreneurship courses began to flourish across US HEIs. Both timelines—the international agenda and the US EE experience—are important for contextualising why and how EE evolved in HEIs in Croatia.

4.1.1 Croatian Political and Economic Context in Which EE Emerged

Croatia emerged from the collapse of the former Yugoslavia in 1991, and this was followed by a war. It was the beginning of a long journey of international recognition:

- 1992: The EU recognized Croatia as an independent state, and Croatia joined the United Nations.
- 1996: Croatia joined the Council of Europe.
- 2009: Croatia joined NATO.
- 2013: 10 years after submitting the application, Croatia joined the European Union.

S. Singer (✉)
UNESCO Chair in Entrepreneurship Education, Josip Juraj Strossmayer University of Osijek, Osijek, Croatia
e-mail: singer@efos.hr

X. Xu, *Comparative Entrepreneurship Education*,
https://doi.org/10.1007/978-981-99-1835-5_4

Croatia is a small country, with a population of 3.8 million in 2021. Its socio-demographic and economic context was built on institutional infrastructure inherited from Yugoslavia by implementing the core principles of the European Union: mobility of people, money, and goods. Besides this exposure to international principles, standards, and programs of institutions that Croatia joined between 1992 and 2013 (Croatia joined the Eurozone in 2023, and is expected to join the OECD), the Croatian context is strongly influenced by two significant changes that began at the same time, in 1991.

The first big change was the transition from the political and economic system that had characterised former Yugoslavia to a new system based on pluralistic democratic principles and the market economy. The challenges of this transition were underestimated and there was not much experience around the world to make it efficient in a short time. The Yugoslav economic system was characterised by a market of goods and services and a limited labor market, no financial market (Vojnić, 1993), self-management, and social ownership over companies (not state-owned). It had been expected that this unique feature of Yugoslav economic system would make the transition toward a market economy easier, but this was not the case. Instead of taking a step forward and enabling the privatization of socially owned companies, the government took a step back by nationalizing those properties and kickstarting the process of privatization, which turned out to be extremely corrupt. Negative feelings among the population accompanied the process from the very beginning, and are still around 30 years later.

The second big change was the 1991–1995 war and the 1996–1998 period of neither-war-nor-peace. Parts of Croatia were occupied and later under UN administration; only at the beginning of 1998 (seven years after the war started) were all territories reintegrated into Croatia. The war caused huge devastation: human, natural, cultural, economic, and educational. Besides the physical devastation, long-lasting devastation of relationships among people and business connections additionally damaged the circumstances in which people lived and worked.

While the change of the political and economic system required a new strategic vision as the highest priority for all the country's stakeholders (people, politicians, educators, businessowners), living on a day-to-day basis through seven years of the war and neither-war-nor-peace pushed it to the back of the waiting list. This resulted in the lack of a strong strategic vision positioning Croatia on the economic map of the European Union. The lack of strategic interventions in building an efficient economic structure is one of the reasons why Croatia had the slowest recovery from the 2008 global financial crisis compared to other EU member states. Only in 2014 did Croatia achieve a positive GDP per capita growth rate. Since 2015, Croatia has maintained a GDP per capita above three percent, which grew even higher in 2021 and 2022, but the regional development gap in Croatia is not closing significantly.

Different international surveys revealed and mutually confirmed several weak points that resulted in Croatia's developmental lag. These include low productivity, low investments in research, and lack of innovative products. According to the European Innovation Scoreboard (European Union, 2021), Croatia belongs in the last

Table 4.1 GNI per capita (Atlas method) (USD)

Country	1998	2021
Croatia	5,470	17,150
Slovenia	11,000	28,240
Ireland	21,430	74,250

Source https://data.worldbank.org/indicator/NY.GNP.PCAP.CD

(fourth) group of Emerging Innovators because its research and innovation performance level is below 70% of the EU average. Slovenia belongs to the third group of Moderate Innovators where performance is between 70 and 100% of the EU average, while Ireland is placed in the second group of Strong Innovators with performance between 100 and 125% of the EU average. Out of 240 regions in the EU, Croatian regions are ranked as 140th, 143rd, 149th, and 187th (European Commission, 2021).

The importance of context can be seen when using GNI per capita to compare changes between 1998 (when all parts of occupied territory were reintegrated into Croatia) and 2021. Table 4.1 compares the GNI per capita for Croatia, Slovenia, and Ireland. The comparison makes it possible to examine differences in the development paths, but these numbers do not reveal the reasons.

All three countries are small according to population size: Croatia, 3.8 million; Slovenia, 2.1 million; and Ireland, 5 million, but the contexts in which these development paths have evolved are different. In 1998, the GNI per capita in Ireland was 3.9 times higher than the GNI per capita in Croatia, and this gap widened in 2021 (4.3). Ireland did not have a war, there was no need to build a new political and economic structure, nor did it experience corrupt privatization. The GNI per capita gap between Croatia and Slovenia shrank (from 2.0 in 1998 to 1.6 in 2021), and here the context plays out differently. Both Slovenia and Croatia emerged from the former Yugoslavia, so they share the same political and economic history. However, Slovenia did not have seven years of war, and therefore should be in a better development position than Croatia. Catching up with Ireland would be more challenging than catching up with Slovenia, but in both cases strategic vision and consistent activities over a longer period of time are crucial ingredients for enacting changes, measured by the people's well-being. The latter is indicated by access to education and health services, equal opportunity for employment and venturing, etc.

4.1.2 Why and How EE Evolved in HEIs

Education is always a part of the solution (or the problem). What the educational system in Croatia did to strengthen the capacity of society to deal with the consequences of changes in the political and economic system, as well as with the consequences of the war devastation, is a legitimate question. EE has a very specific role in all this because it brings new knowledge and skills that can be defined from the broader (mindset) or narrower perspective (recognizing opportunity and starting a venture).

It will be shown again that context matters, and that this is not only the case in Croatia. The Harvard Business School is a good example of how context provoked EE, starting the first entrepreneurship course in 1947 in response to the needs of World War II veterans who wanted to start their own ventures (HBS Alumni Bulletin, 1996). A different context triggered EE in Croatia: jobs were lost due to a devastated economy (as a result of the war and corrupt privatization), but also due to the changed political and economic system.

In former Yugoslavia, business education was either neutral or explicitly focused on the management of socially owned big companies. Starting and developing small private businesses (e.g., agriculture, crafts, services, manufacturing) were not the focus of education. Therefore, there had been an immediate need to redesign business education, but this did not happen.

There were no anticipatory strategies and activities to minimize the consequences of the mega changes Croatia was faced with—only reactions, on the policy and individual levels.

Policy reactions to the mega changes—building a new political and economic system and the war—were slow, late, mutually disconnected, and mostly triggered by the process of becoming a member state of the EU. The situation has not changed much since. The first strategic document on entrepreneurship learning for 2010–2014 (Government of Croatia, 2010) was not even mentioned in the strategy document on the development of entrepreneurship for 2013–2020 (Ministry of Entrepreneurship & Crafts, 2013). The latter document identified the following strategic goals: improve the innovative capacity of SMEs; develop new financial instruments for SMEs; develop entrepreneurship ecosystems in order to achieve a more balanced regional development; strengthen entrepreneurial skills (lifelong learning); and improve the business environment (by eliminating administrative barriers). Although it was planned to evaluate the achievements in preparation for a new strategy after 2000, there was neither an evaluation report nor a new strategic document. Both strategic documents were developed by the Ministry of Entrepreneurship without collaboration from the Ministry of Science and Education, which thus reduced implementation capacity and the probability of achieving goals. This was especially relevant for the strategy on learning for entrepreneurship, where both identified goals—to develop a positive attitude toward lifelong learning; and to include entrepreneurship competence in all types and levels of formal, nonformal and informal education and learning—require collaboration and involvement with the Ministry of Science and Education. It was planned that the implementation of this strategy over a period of five years would increase employability, self-employment, and the number of startups based on opportunity recognition and not on necessity. Although the document had a timeline, there were no evaluation reports, nor was the strategy renewed.

At the individual level, responses to mega changes were quicker, with some attempts to open anticipatory 'if' questions, but usually without success.[1] The war triggered a reaction from nongovernmental organisations, using the broad definition of human rights, including employment issues. At the same time, the change of the political and economic system, which had been pushed aside by the war, required a serious, overarching overhaul, with the priority of redesigning higher education because of its transversal feature and long-term influence on society. This did not happen at the national strategic and policy level, but was left to initiatives by institutions or individuals. The very first attempt to introduce EE was the creation of an undergraduate program in entrepreneurship in 1990, based on similar programs in the US. But it was too late: Yugoslavia disintegrated in 1991 and the war started. No one cared about introducing a new educational programme; academic year 1991/1992 was cancelled in eastern Croatia, where the Josip Juraj (J.J.) Strossmayer University of Osijek is located, and all primary and secondary schools were displaced either to western Croatia or to neighbouring countries like Hungary and Czechoslovakia.

The second attempt to introduce EE originated from a group of researchers/educators at J.J. Strossmayer University of Osijek in 1996. Again, context is important for understanding why researchers from that particular university came up with such an initiative. In the late 1980s, there was some political discussion on redefining the Yugoslav economic system by combining the concept of social ownership and workers' self-management with market economy, which would make the Yugoslav economy more innovative and competitive. Several researchers from this first initiative were actively researching the weak points of the Yugoslav economy, especially how to identify value creation in managing development processes (Singer et al., 1987–1990). When everything changed with the new political and economic system, and when war broke out, this group of researchers/educators asked 'if' questions: where would Croatia go IF education stayed 'business as usual'; could the development path be redesigned toward the catching-up mode IF education equipped young people with knowledge and the skills of being proactive, innovative, and responsible for one's own choices?

In order to find the answers, the researchers/educators, who were the only research group in Croatia exclusively focused on entrepreneurship, investigated changes in education in the US, one of the most developed and dynamic economies in the world. They used the US experience of introducing entrepreneurial competences in education so as to build a vibrant entrepreneurial economy (Research projects on entrepreneurship, from 1991).

The group's research efforts led to two major findings that supported their intention to introduce EE in Croatia: the entrepreneurship phenomenon is receiving global attention from researchers, and has entered the field of education. The rising number of educational programs in entrepreneurship in the US provoked the question: if

[1] During the war years (1991–1995), I was active in some human rights activities and had opportunities to discuss 'if' questions, trying to anticipate activities *after* the war with some foreign diplomats, including UN officials, who were visiting the eastern, occupied part of Croatia. The answers were usually "it is too early to think about it, that it is time to think about humanitarian aid."

Americans need EE, what about Croatia? Osijek was on the edge of occupied territory, most companies were displaced to other parts of Croatia or to neighbouring countries—Osijek's population shrank from 140,000 in the late 1980s to about 10,000 in 1991–1992, and then gradually increased when people started returning. In 2021, the population reached 96,848 (Croatian Bureau of Statistics, 2021).

In this context, the second attempt to introduce EE at the J.J. Strossmayer University of Osijek in 1996 was driven by a reason different from the first: it stemmed from the awareness that many people had lost their jobs because of the war devastation and corrupt privatization. From researching entrepreneurship, this group of researchers learnt that introducing a new program, based only on US best practices, would not solve the problems in Croatia if it was not contextualised, accompanied with an own ecosystem and trained teachers.[2] The new approach tried to combine an immediate reaction to devastation and anticipate needs for the economy's recovery by simultaneously engaging in two clusters of activities:

a. Developing an entrepreneurship ecosystem (access to money, access to skills for starting and growing a business venture, a think-tank focused on small businesses and entrepreneurship).
b. Developing a new research-based university program focused on building entrepreneurial skills among young people.

At the same time, the HBS Alumni Bulletin (December 1, 1996) presented "Entrepreneurship at HBS," a message from Dean Clark. It referred to the beginning of EE at the HBS, which had begun in 1947. Three sentences triggered the attention of this group of researchers in war-devastated eastern Croatia. The first sentence emphasized how the new course, "Management of New Enterprises," offered in 1947 "served the needs of World War II GIs eager to launch entrepreneurial ventures and make their mark in the business world" and was "a significant departure from more traditional classes aimed at training managers to lead large, complex corporation." The third sentence, "business leaders in large and small companies alike now need to be versed not only in running the established enterprise but also in entrepreneurial action, which embraces change and seizes opportunities," confirmed the approach to designing the EE program in Croatia, which had been introduced at the J.J. Strossmayer University of Osijek in 2000. The Croatian context included the dimension of the changed political and economic system and how business education had to be redesigned by focusing on small business initiatives, not only on the management of big, established companies.

The process of introducing EE in Croatia, through programs at the J.J. Strossmayer University of Osijek, has a strong footprint of Allan Gibb's work on EE, entrepreneurial university, and the entrepreneurship ecosystem (Gibb, 1993a, 1993b,

[2] The first initiative from 1990 was quite naive, and probably would not be successful because it was more or less a result of copying the US experience without contextualisation and without preparing the faculty for delivering such a new program.

2000a, 2000b, 2002, 2005; Gibb & Haskins, 2013).[3] Gibb's approach to understanding the world of the owner-manager was the starting point for Osijek's initiative to develop an entrepreneurship ecosystem centered on the needs of the owner-manager. Gibb always had a contextualised lens and emphasized the importance of place and cultural context. His major influence on EE in Croatia is his view on impact, which he saw as a spirally intertwined path of grounding the research in practice and then using the findings to design and test educational programs for delivery.

The EE process established at the J.J. Strossmayer University of Osijek had a palpable spill-over effect throughout Croatia. EE in Osijek started in 2000 with a postgraduate specialist program; an undergraduate program was introduced in 2005, and the doctoral programme, Entrepreneurship and Innovativeness, enrolled its first cohort in 2010. Some doctoral students joined other HEIs in Croatia, where they started courses on entrepreneurship, and many postgraduates were engaged in developing the entrepreneurship ecosystem across Croatia, such as incubators and centers for entrepreneurship. Because of this, J.J. Strossmayer University of Osijek is recognized as a nurturing place for researchers, teaching them how to investigate entrepreneurship phenomena and find evidence on why entrepreneurship competences in the broadest sense—to be proactive, innovative, and responsible for one's choices—are relevant and important in the development process at the individual and institutional levels. This pioneering work triggered the interest of other HEIs in Croatia, which then introduced their own EE programs at the undergraduate or graduate levels.

Croatia's experience in starting EE under specific circumstances was presented as one of 20 cases in the survey *Supporting the Entrepreneurial Potential of Higher Education* (European Commission, 2015, pp. 364–390). Osijek's efforts were also studied by Harmeling et al. (2004) and presented at the Kauffman Foundation's *Entrepreneurship Research Conference: Frontiers of Entrepreneurship Research* in 2004.

4.2 Organisational Structure and Culture of HEIs: Supporting or Hindering EE

The basic principles of education in Croatia are defined by the Constitution of the Republic of Croatia (Articles 65–66). Primary education is free and compulsory, while secondary and higher education is available to everyone under equal conditions and in accordance with the strategic vision and capacity including resources, organisation, and culture. HEIs are organized as universities, polytechnics, and colleges. These institutions are either public or private, fulfilling the same legal requirements of the Act on Scientific Activity and Higher Education (Official Gazette, 123/2003).

---

[3] Allan Gibb was a member, mentor, and educator in the establishment of EE programs at the J.J. Strossmayer University of Osijek from 1996 to 2018.

All higher education study programs in Croatia are aligned with the Bologna Process in 2005, which established the European Higher Education Area.

4.2.1 Strategic and Legal Framework of Higher Education in Croatia

Within the framework of higher education, there are some differences between universities and polytechnics/colleges. Universities organize and implement university study programs, but they also organize and implement professional study programs. Polytechnics and colleges are legally entitled to carry out higher education activities in the form of organising and implementing professional study programs.

University study programs comprise three levels: undergraduate, graduate, and postgraduate:

- Undergraduate programs typically last three (180 ECTS[4]) to four (240 ECTS) years. Upon completion, students are awarded a Bachelor's degree with a specialisation.
- Graduate programs typically last one (60 ECTS) to two (120 ECTS) years. The total number of credits earned after completing both undergraduate and graduate studies is at least 300 ECTS. Upon completion of both undergraduate and graduate studies, students are awarded a Master's degree with a specialisation.
- Postgraduate university studies are divided into specialist and doctoral studies. Specialist studies last one to two years. Upon completion, the student is awarded the title of University Specialist with a specialisation.
- Doctoral studies last three years (180 ECTS). Upon completion, the academic title of Doctor of Science or Doctor of Arts is awarded.

Professional studies are divided into short professional studies, undergraduate professional studies, and specialist professional graduate studies:

- Short professional studies last two (120 ECTS) to two-and-a-half (150 ECTS) years. Upon completion, students are awarded a professional title with a specialisation.
- Undergraduate professional studies last three (sometimes four) years and students earn 180–240 ECTS. Upon completion, students are awarded a professional title with a specialization.
- Specialist professional graduate studies last one to two years with 60–120 ECTS. Upon completion, students are awarded the title of a specialist of the respective profession.

The total number of credits earned after completing both undergraduate and graduate professional studies is at least 300 ECTS.

---

[4] European Credit Transfer and Accumulation System is a tool of the European Higher Education Area for making studies and courses more transparent.

The number of students fluctuates between 166,061 in 2013/2014 and 161,627 in 2019/2020. Majority of the students (90%) study at public HEIs, and 82% are enrolled in university-based study programs (Agency for Science and Education, 2022).

The strategic framework for public HEIs' function is defined by several documents (Strategy of Education, Science and Technology—New Colours of Knowledge, 2014; Strategic Plans of the Ministry of Science and Education, 2018–2020, 2020–2022), which seek to improve the employability of students, increase the number of STEM students, build a more supportive learning ecosystem, and improve the quality of doctoral programs by strengthening the criteria for accreditation.

In the strategic plans of 2018–2020 and 2020–2022, the Ministry of Science and Education specifically refers to the need for including entrepreneurship competence in education in order to "produce" competent, enterprising, and responsible people. The strategic plan 2020–2022 also refers to the need for developing the entrepreneurship eco-system by establishing technology centers and business incubators, etc., which will contribute to intensive collaboration between universities and the business sector, making knowledge transfer more effective and efficient.

Analysis of the strategic framework indicates some restraint in envisioning the role of HEIs in the overall development of Croatia. It was also observed in the European Training Foundation's (ETF) analysis of human resource development in Croatia (Rinaldi et al., 2012, p. 31) that "cooperation between higher education institutions, research and technological centres and enterprises is neither analysed nor regulated at national level. Many universities still lack entrepreneurial spirit and are too academically oriented." It should not contradict the university's autonomy to have a jointly developed strategic vision for the future of education.

The autonomy of universities in organising these activities is guaranteed by Article 67 of the Constitution. Universities independently decide on their organisation and operations in compliance with legal requirements stipulated by the Act on Scientific Activity and Higher Education (Official Gazette, 123/2003). The governing body of the university is the senate, which is in charge of the most important academic, research, educational, and financial issues of the university. The rector manages the university and chairs the senate.

4.2.2 Strategic Vitality and Organisational Capability of HEIs Relating to Entrepreneurship Agenda—Through the Lens of International Reviews

Using external reviews of the Croatian higher education system, with a focus on entrepreneurship competences, it is possible to identify the process of change. Four such audits are used:

- ETF (2008): human resource development, country analysis (Skjolstrup, 2008).
- ETF (2012): human resource development, country analysis (Rinaldi et al., 2012).

- European Commission (2016): *Supporting the Entrepreneurial Potential of Higher Education, Final Report* (European Commission et al., 2016).
- OECD/European Commission (2019): *Supporting Entrepreneurship and Innovation in Higher Education Croatia.*

In 2008, the ETF analysis suggested that "the long-term ought to be to move towards the key competences identified in the EU Reference Framework for Key Competences" (Skjolstrup, 2008, p. 15), including entrepreneurship. It is interesting that management training is "a particularly important area in Croatia" (p. 17), because private companies operating in a market economy require different knowledge and skills than in the previous industrial structure. The analysis confirmed positive changes in management training for big companies, but indicated problems in providing such training for managers in SMEs, and that continuous learning is "not taken sufficiently seriously" (pp. 17–18). A couple of recommendations followed the analysis:

- "To adjust educational programmes to the knowledge and skills needed in the future, with a particular emphasis on the development of interdisciplinary and key competences at all levels of education (particularly, the entrepreneurial way of thinking)" (p. 32).
- To establish centres for management education and training at the regional level for "offering training in entrepreneurship to encourage the creation of new small businesses" (p. 18).

In 2012, the ETF analysis listed two institutional initiatives focused on education and training for entrepreneurship: the Education for Entrepreneurship (E4E) and the South East European Centre for Entrepreneurial Learning (SEECEL), co-founded by the Croatian government and supported by EU funds. Neither of these have sustained their activities. The ETF analysis confirmed a low level of awareness in the business community of benefits from having a closer relationship with the education sector and investing in human resources development, although entrepreneurship promotion has been very good in Croatia (OECD et al., 2009). An important observation relates to the national Strategy of Entrepreneurial Learning 2010–2014 adopted in July 2010: "The challenge now is to implement this strategy" (Rinaldi et al., 2012, p. 27). The strategy was not renewed, and there were no reports on implementation results. Rinaldi et al. (2012) reported more projects focusing on entrepreneurship and funded through different EU schemes. They recommended introducing EE as a cross-curricular principle in HEIs.

In 2016, the European Commission performed an analysis of EE in Europe, based on 20 examples selected with following criteria: new EE models; active for a longer period; different aspects of EE (curricular, extracurricular, institutional design); different types of universities; and from different countries. The cases were from 19 EU member states. From Croatia, the J.J. Strossmayer University of Osijek was selected due to its pioneering role in promoting university-based EE. The survey covered three main issues of EE: curricular offerings, extracurricular activities, and institutional aspects such as EE design, delivery modes, organisational setting, and legal framework. The data collected also included information on the influence of sociocultural, economic, and political contexts, as well as the impact of EE on society and economy. The major findings identified the following impediments, along with possible solutions (European Commission et al., 2016, p. 8):

Challenge 1: Overcoming reservations against EE on the part of university managers, educators, and students. One solution is to see EE not only as teaching to start a new business but also, more broadly, creatively making an idea happen.
Challenge 2: Assuring sustainable finance for EE as a relatively young and personnel-intensive discipline. One solution is to offer EE as paid education to professionals.
Challenge 3: Assuring curricular EE quality when experience is limited, with new methods arising, and when leading educators leave. "Educating the educators" as well as national and international entrepreneurship educator networks may help. If legal framework conditions for EE are unfavourable, e.g., involving practitioners in teaching, lobbying for changes can be considered.
Challenge 4: Assuring sustainability and quality of extracurricular activities to maintain their flexibility while improving their institutionalisation.
Challenge 5: Assuring strong networks for supporting EE, primarily with alumni who can be guest speakers, mentors, and funders.
Challenge 6: Measuring outcomes and impact of EE with a long-term perspective. Assessment should be focused not only on startups, but also on students' entrepreneurial mindsets, skills, and behavior before and after courses.

The survey confirmed that there is no one-size-fits-all solution because universities conduct their activities in very different socioeconomic, political, legal, and cultural contexts. Also, the survey indicated the problem of analysing EE when it is isolated from research, not taking into account the university structure and organisational culture, which can be a strong internal impediment.

In 2019, in partnership with the European Commission, the OECD performed a review of the impact of HEIs on entrepreneurship and innovation in Croatia, using the HEInnovate[5] assessment approach. The HEIs used as case studies included a combination of public and private HEIs, universities and polytechnics, located across Croatia. The report emphasizes two important aspects of higher education in Croatia: the lack of strategic vision, and "a peculiar governance structure compared with other international higher education systems" (OECD/European Commission, 2019, p. 13). Public universities either have a centralised governance structure (the rector and the senate have governing power) or a "non-integrated structure" (faculties are independent legal entities with significant administrative and financial autonomy from university management).

Key findings warn about lagging behind in the entrepreneurial and innovation agenda of HEIs:

[5] HEInnovate is a holistic guiding framework for supporting innovation and entrepreneurship in higher education, developed by the OECD and the European Commission. HEInnovate offers an online self-assessment tool for HEIs (www.heinnovate.eu), available in 24 languages.

- While all HEIs involved in this assessment have developed different activities (joint research projects, internships, mentoring programs, etc.) to promote EE, there are some limits in the capacity of the system to mainstream new innovative practices.
- EE is still connected to economics/business schools; there are no organisational arrangements to enable students to access EE across disciplines. Some of assessed HEIs do not have EE as a priority in their mission statements.
- The participation in EU programs promoting education and innovation contributed to building the capacity of HEIs for a proactive role in their own ecosystems. But due to a lack of resources to support such activities (e.g., technology parks, business incubators, centers for entrepreneurship), the systemic capacity of university to engage with stakeholders is limited.
- Outdated regulation on intellectual property rights and legal obstacles to recruiting staff from non-academic backgrounds additionally limit HEIs' capacity to collaborate with external stakeholders.
- Innovative and entrepreneurial initiatives are implemented on an ad-hoc basis, mostly based on the enthusiasm and commitment of individual faculty, with little institutional support or recognition. Strong autonomy at the faculty level, underdeveloped managerial capacity in HEIs, the absence of prioritizing EE, and sporadic availability for innovation and entrepreneurship, are major causes of a lack of systemic structure needed to introduce changes in HEIs.

Based on such findings, the HEInnovate report provided systemic recommendations to the national government and granular recommendations to HEIs, focused on the most challenging factors that impede the efficient implementation of innovation and entrepreneurship strategies:

- For the national government: (1) to establish an efficient higher education funding system based on a transparent linkage between strategic objectives and funding; (2) to change the regulatory framework restricting the capacity of engaging faculty members in public universities in knowledge-exchange activities with businesses, nonprofit organisations, government agencies, and other HEIs; (3) to continue reforming the governance of public universities toward becoming integrated, because the present situation of dis-integrated universities leads to many fragmented strategies promoting entrepreneurial and innovation agenda without coordinating mechanisms.
- For HEIs: (1) to mainstream EE across campus by developing courses across faculties, or general courses with specific faculty-related modules; (2) to create centers for entrepreneurship and innovation at the university level to avoid resource fragmentation and to facilitate interaction between HEIs and the business community, framed in a common and coherent institutional strategy; (3) to recognize knowledge-exchange activities as part of academic reward and promotion criteria for faculty members.

All the presented reviews covering 2008–2019 affirm each other, adding new insights on the strategic and legal framework. If the framework is not a direct obstacle, neither does it support HEIs' institutional capacity building to develop and implement EE.

4.2.3 *Case of EE at the J.J. Strossmayer University of Osijek: Developing EE from Scratch*

The J.J. Strossmayer University of Osijek is a public university established in 1975, with 18,000 students and 2,100 faculty and non-academic staff (2019). The academic structure comprises 17 faculties and departments and seven units with a supporting role as legally independent units. In the rector's 2018/2019 report, this non-integrated feature of the university was neglected. It also claimed that so-called functional integration was in place, through the university's development strategy (Sveučilište Josipa Jurja Strossmayera u Osijeku, 2020, p. 1). The university's 2021–2030 strategy (Sveučilište Josipa Jurja Strossmayera u Osijeku, 2021) stipulates two goals that could have an integrative nature:

- Promote excellence in study programs through strengthening educators' competences and development tools for learning in digital environments and by including entrepreneurship in study programs (p. 3).
- Establish a unit for collaboration with the local community, which would improve strategic partnerships in education, research, innovation, and entrepreneurship (pp. 47–48).

There is no top management position relating to EE in the university's hierarchy.

The initiative of a group of researchers/educators to start a study program focused on entrepreneurship competences in the late 1990s is a good example of navigating the maze of hindrances relating to internal organisational structure and culture, which delayed the first enrolment until 2000. Reactions to the devastation caused by the 1991–1995 war and corrupt privatization shaped the context in which this initiative was launched. The jobs that were lost could not be recovered through the remaining businesses; new energy was needed for self-employment and for starting businesses with growth potential. Starting a university-based EE program was not the first move of this research group—it came after learning some lessons from other initiatives that had been launched earlier in 1995–1996. The first response to the postwar situation was to provide financial access for starting new ventures, and the first micro-finance institution was established in Osijek in 1996. Soon, it was obvious that money without knowledge and skills was not enough, which led to the next step of providing training and mentoring support for people interested in self-employment or starting their own business. Therefore, in 1997 they established the Center for Entrepreneurship in Osijek. Only then did this group of researchers decide to start educating young people on entrepreneurial competences through formal university education. This decision was made in a context that can be described with a statement of the leader of this initiative: "We did not have the money; we did not have people." They started with what they had.

4.2.3.1 Chronology

2000: Launch of graduate program (Master of Science) in entrepreneurship at the Faculty of Economics.

As the initiators of the EE program had been intensively engaged in researching entrepreneurship since the beginning of 1990s, with some research activities conducted even in the late 1980s, they decided to start the graduate entrepreneurship program (Master of Science), and not an undergraduate program like the first attempt in 1990. The reason for the change in the focus was the long-term vision to develop a program based on combined local and international faculty, instead of depending only on outside experts. The master's program was used as the first step toward fulfilling this vision because it was expected that some graduates would be interested in doctoral studies. There were internal and external impediments to starting the program. Although the group behind the initiative designed the program for implementation across university, using the experience of the Center for Entrepreneurial Learning at the University of Cambridge as inspiration, this was not possible: university management was not interested as they perceived that entrepreneurs were born and "street-smart," and did not see it as a topic that needed to be taught. The solution was to place it in the portfolio of educational programs of the Faculty of Economics, because all the initiators were from that unit. But even there, some faculty members questioned the initiative: "Why do we need entrepreneurship when we have management?"

An external impediment came from the Ministry of Science and Education. It took them two years to approve the program after the proposal submission in 1998.The National Council for Higher Education in Croatia refused the proposal in June 1999, citing various reasons: the concept of the program focused on SMEs and ignoring large companies; the use of mostly foreign literature instead of publications by professors from Osijek; and the University of Osijek was considered too small to handle the program. After a year of contending with the Ministry, when proponents of the initiative provided many international examples of such programs, the Ministry of Science and Education approved the program in May 2000. One week after the ministry's approval, the entrepreneurial Master's program enrolled its first students. This rapid start was because a group of students had been waiting for around two years for the program to begin.

2005: The undergraduate EE program was introduced, designed along the principles of the Bologna Process, which had been implemented to establish the European Higher Education Area. Graduate and postgraduate specialist programs in entrepreneurship were also redesigned according to Bologna principles.

2010: The postgraduate doctoral program, Entrepreneurship and Innovativeness, was introduced, not within the Faculty of Economics, but as part of the International Centre for Entrepreneurial Studies (ICES). ICES was established as a transversal university unit to house this doctoral program, as well as potentially all other EE programs and research activities. Both the doctoral program and ICES came about as a joint outcome of the European Union's TEMPUS (Trans-European Mobility Programme for University Studies) 2007–2010 funding program in the framework of a joint effort of five universities: Osijek, Croatia; Turku, Finland; Durham, UK; Klagenfurt, Austria; and Maribor, Slovenia). The program was delivered in English and in hybrid mode (physical and online). The enrollment and faculty were international, including countries such as Canada, US, China, Nigeria, Greece, Kosovo, Bosnia and Herzegovina, Finland, the Netherlands, Slovenia, and so on. Students came from different disciplines such as engineering, economics, law, agriculture, and humanities, which attests to the interdisciplinary nature of the program.

4.2.3.2 Key Perspectives of Developing EE Programs

EE programs were developed based on systems theory. Concepts included interconnectedness, openness, equifinality (Ludwig von Bertalanffy and Kenneth Boulding), effectuation (Saras Sarasvathy), innovation (Joseph Alois Schumpeter), the owner-manager perspective (Allan Gibb), and entrepreneurship (Howard Stevenson, Allan Gibb and Jerome Katz).

4.2.3.3 Content

All EE programs aim to develop an entrepreneurial mindset in the learning process, based on enterprising competences being proactiveness, innovativeness in defining and solving problems, and the capacity to be responsible for one's own choices. Acquiring the knowledge and skills to start a business and understanding the life cycle of a venture are also part of the content.

4.2.3.4 Learning Methodology

In order to achieve an efficient and effective learning process, it is important to depart from the traditional curriculum approach, i.e., courses and semesters. The initial design of educational programs, especially the doctoral program Entrepreneurship and Innovativeness, focused on project-based learning. However, due to legal provisions relating to the curricula, as well as the university organisation and culture, this could not be implemented. Therefore, the form of the curricula remained, in terms of having courses and semesters, but the learning methodology emphasizes project-based learning, team teaching, roleplaying, drama, practical cases, experiential learning, and gaining experience from guest speakers and from practice. The aim is to expose students to real, practical problems, not only pertaining to entrepreneurship, but also in a wider scope such as discussing problems in the classroom and with practitioners. For example, in business ethics, students deal with real examples of ethical and unethical behavior, drawn from Croatia or from companies around the world. These learning methods require identifying problems and solutions, and teachers become mentors in this process. However, two important impediments emerged: teachers lack knowledge and skills to become mentors, and there are legal difficulties bringing professors of practice to the classroom. For example, a person from outside the university can be invited as a guest speaker, but cannot be in charge of the whole course without a PhD degree. This makes the combination of academia and practice much more difficult.

4.2.3.5 Human Resources

To impart quality learning methods requires mentoring skills, the recognition of problems in the community or globally, and creative problem-solving. The ICES management thus implemented teacher-training programs, either on-site or sending them abroad. For example, the model of team teaching was adopted from Prof. Jerome Katz from St. Louis University. The use of cases in teaching and the roleplaying method were picked up at the Harvard Business School at teacher-training events. Below are some examples at the beginning of the program:

1998–2000: Consultancy training (Piotr Korynski, Open Society Institute, New York, Director of the Economic Development Programme) in Osijek.

2001: Case study teaching (Susan Harmeling, professional case writer, Harvard Business School) in Osijek.

2002: "Train the Trainer" workshops on how to design a course as well as identify course outcomes and expected competences (Joan Gillman, University of Wisconsin Business School, US, and Deborah Laurel, independent consultant) in Osijek.

2004: Hybrid training in proactiveness and initiative in "authentic leadership" (Susan Skjei and Barbara Lawton, Naropa University, US) in Osijek.

Since 2004 (yearly basis): Case study approach in "Microeconomics of Competitiveness" (Prof. Michael Porter, Harvard Business School) in Boston, US.

2005: "Experiential Classroom," a three-day training program for lecturers in entrepreneurship at Syracuse University, US.

2008: "European Entrepreneurship Colloquium for Participant-centered Learning (EECPCL)" at Harvard Business School, Boston, US.

2010–2012: European Entrepreneurship Education Summer School comprising three workshops at the University of Turku, Finland; Aarhus Business School, Denmark; and J.J. Strossmayer University of Osijek, Croatia.

The "shadowing" method is another approach to develop staff competency. Younger faculty members were assigned to experienced external lecturers, building up their competency by visiting classes and working together with external lecturers.

With these approaches, it was possible to develop a strong group of locals who could, teaming up with professors from abroad, implement the program by departing from traditional teaching.

4.2.3.6 Financial Resources

The university started its entrepreneurship activities with very limited human and financial resources. The university received startup funds of USD200,000 from the Open Society Foundation in New York. This support was necessary, especially for bringing foreign lecturers into the program. However, since funding was still limited, many professors came pro-bono appreciating commitment of local group engaged in this initiative and sometimes did not even ask for reimbursement of traveling costs. This was another essential support that allowed the program to develop.

4.2.3.7 International Networking

EE programs were developed as a response to the economic devastation of eastern Croatia, with the strong support of a very broad international network of experts. The advisory board, chaired by Prof. Howard Stevenson from the Harvard Business School offered a strategic vision for the programs. Prof. Allan Gibb had been part of the initiative for EE programmes since 1996; he mentored, taught, advised, and broadened the network of internationally recognised experts on entrepreneurship and small businesses (e.g., Jerome Katz [US], Antti Paasio [Finland], and David Pistrui [US]).

4.2.3.8 Extracurricular Activities

There are four major extracurricular EE activities: community work, the "Entrepreneurs Without Borders" initiative, consulting work, and a business plan competition.

For her contributions to developing EE in Croatia, to constructive collaboration in international projects on EE, and to entrepreneurship research, the head of the EE programs, Slavica Singer, was awarded the UNESCO Chair in Entrepreneurship Education in 2008.

4.3 Entrepreneurial Environment and Ecosystem: Supporting or Hindering EE at HEIs

There is no way to learn without interacting with different actors (people, institutions, nature, the planet). These interactions may be observing, talking, reading, or experiencing, and they may happen in a broader context (the environment) or in a more immediate context (the ecosystem). Awareness of the environment or ecosystem in which learning/education occurs evolves from reactions to anticipations, from the fragmented recognition of some components and isolated interactions to their holistic features.

4.3.1 Systems Approach in Designing a Supportive Context for EE

The difference between *environment* and *ecosystem* brings a subtle but important perspective in defining the context in which EE is organized and implemented. The ecosystem includes components within the immediate reach of many, or whose quality influences many people who are usually connected to regional policies. The environment is defined more globally, with varied influences on those who are usually connected to national-level policies.[6] Indirectly focusing on the definition of the entrepreneurship ecosystem, Isenberg (2010) provides a clear explanation of the differences between these two concepts. Various components in the environment are individually conducive to entrepreneurship, but without effective interrelations among those components, they are insufficient to sustain it. Therefore, the architecture of the context based on the systems approach is essential for understanding the hindering or supportive feature of the context in which EE is organized and implemented. Pfeifer et al. (2021, p. 2) emphasize the spatial dimension of the influence for differentiating between the national and sub-national contexts. The national level of influence covers stakeholders' activities that horizontally affect all actors in a country (e.g., government policies on regulatory frameworks). The sub-national level, which is usually described as "entrepreneurship ecosystem" in recent scholarly discussions, refers to the availability of conditions more specific to entrepreneurship activity in a region (e.g., educational infrastructure in a region, or the availability of services of an incubator/accelerator, etc.).

In this Section 4.3, the difference between two contexts (environment and ecosystem) will be labelled as *ecosystem on the national level* (for environment) and *ecosystem on the sub-national level,* otherwise *ecosystem*, if the differences are not essential.

The interrelations that classify a context as an ecosystem can be clustered as:

- Interrelations IN the ecosystem—among different components of the ecosystem.
- Interrelations BETWEEN the ecosystem as a whole or its components, AND the learner.

Labelling such interactions as an ecosystem implies a holistic approach. It is in line with the longstanding philosophical discussion on capturing the truth ("the truth is the whole," Phen. §20, and that truth is "actual" only to the extent that it achieves the

[6] The wording of *environment* and *ecosystem* would be less important if it does not have implications on identification of stakeholders and their responsibilities for building a supportive entrepreneurial ecosystem. The word *system* emphasizes a complex whole in which interrelations are recognized and could be managed more efficiently than in the broader context of the environment.

form of the "system," Phen. §25, Hegel, 2000). By adding the word *entrepreneurial*, Stam (2015, p. 2) refers to entrepreneurship, a process in which opportunities for creating new goods and services are explored, evaluated, and exploited (Shane & Venkataraman, 2000), or by which individuals exploit opportunities for innovation (Schumpeter, 1934).

Efficient design in terms of content, organisation, delivery, and assessment, as well as the implementation of EE at HEIs, requires a supportive ecosystem at both national and sub-national levels.

4.3.2 Entrepreneurial Ecosystem at the National Level (Environment) and EE

At the national level, the entrepreneurial ecosystem includes some components that require government interventions such as policies, programs, regulations, taxes, intellectual property, and education. These are not in the immediate reach of sub-national institutions. Depending on the political and cultural contexts, the entrepreneurial ecosystem at the national level can differ in the domain composition. For example, Isenberg (2011, pp. 6–7) emphasizes the uniqueness of the combinations of major domains of any entrepreneurship ecosystem: conducive policy, finance, culture, support, human capital, and markets, without identifying causal paths. This is an important feature of Isenberg's approach because it places holistic thinking behind designing a conducive and sustained entrepreneurship ecosystem.

On the policy level, there are only a few policy documents relating to EE that directly refer to the entrepreneurial ecosystem. At the time of this analysis (2022), only two documents cover the 2021–2030 period, with marginal implications on the context in which EE is conducted. The *National Development Strategy of the Republic of Croatia*, which covers till 2030 (Government of Croatia, 2021a), only loosely identified promoting lifelong learning, self-employment, and gaining entrepreneurial competences through education. The low level of entrepreneurial culture in less developed regions and the need to develop regional ecosystems to support entrepreneurs are only mentioned as issues. The *Recovery and Resilience Plan for Croatia 2021–2026* (Government of Croatia, 2021b) introduces activities relating to strengthening entrepreneurial skills through education as well as building a supportive entrepreneurship ecosystem at the national level, specifically for innovative ventures.

The most comprehensive insight into the quality of national-level entrepreneurial ecosystem is provided by the Global Entrepreneurship Monitor (GEM)[7] survey, the world's biggest entrepreneurship survey since 2000. GEM conceptually anticipates that entrepreneurial activities result from the interactions of individuals with a specific set of attributes and social values, with components of the entrepreneurial framework

[7] https://www.gemconsortium.org.

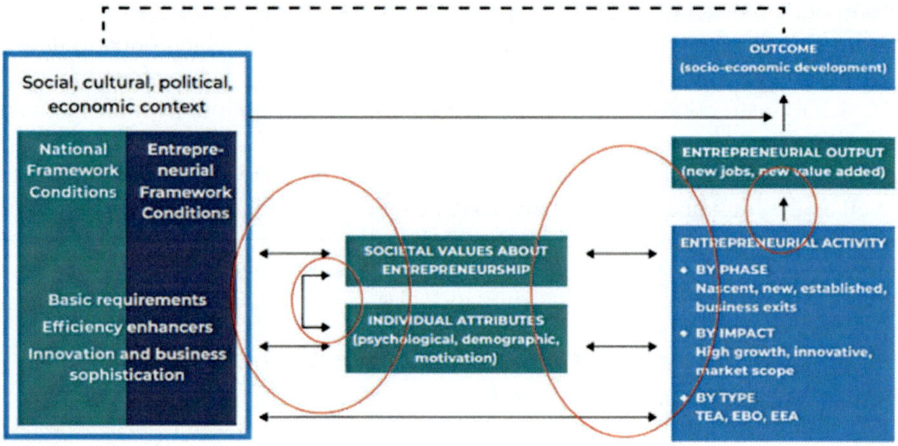

Fig. 4.1 Global Entrepreneurship Monitor conceptual framework. *Source* GEM (Global Entrepreneurship Monitor) (2022). Global Entrepreneurship Monitor 2021/2022 Global Report: Opportunity Amid Disruption. London: GEM

conditions (national-level entrepreneurial ecosystem) in which such individuals act (see Fig. 4.1).

GEM's Entrepreneurial Framework Conditions include components that represent the social, cultural, political, and economic contexts:

- entrepreneurial finance;
- ease of access to entrepreneurial finance;
- government policy (its support and relevance as well as taxes and bureaucracy);
- government entrepreneurship programs;
- EE (EE in schools, colleges, and universities);
- research and development transfer;
- commercial and professional infrastructure;
- ease of entry (market dynamics as well as market burdens and regulations);
- physical infrastructure; and
- social and cultural norms.

GEM evaluates the national-level entrepreneurial ecosystem by using the NECI (National Entrepreneurship Context Index) composite index and individual scores for each component. All indicators are based on experts' evaluations using an 11-point Likert scale, ranging from 0 to 10.[8] Score 5 can be regarded as a divide between sufficient and less sufficient scores for a specific indicator. The systemic feature of the GEM Entrepreneurial Framework Conditions—the interconnectedness of domains—enables the identification of bottleneck domains or components that limit the supportive capacity of the entrepreneurial environment as a whole.

[8] Scale values: 0 = very inadequate insufficient status, 10 = very adequate sufficient status.

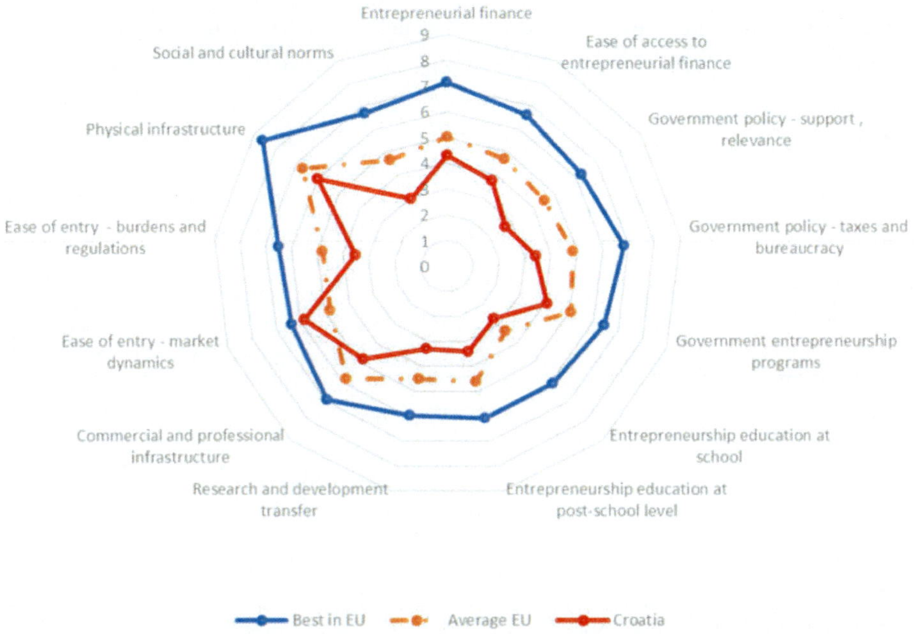

Fig. 4.2 Expert ratings of entrepreneurial framework conditions, GEM 2021. *Source* Singer et al. (2022)

The Croatian national-level entrepreneurial ecosystem has only two out of 13 components with scores higher than 5: Market Dynamics (5.83) and Physical Infrastructure (6.0). EE at the post-school level is rated lowest (3.41) among European Union countries in 2020 and 2021.[9] In addition, the domain of Social and Cultural Norms has had the lowest scores for years (2.96 in 2021; especially the following statements: "*the national culture is highly supportive of individual success achieved through own personal efforts*"; "*the national culture encourages entrepreneurial risk-taking*") (Singer et al., 2022).

A comparison between Croatia and the highest-rated domains of national-level entrepreneurial ecosystems in EU countries that participated in the 2021 survey indicates its developmental lag from the best-performing countries, and allows researchers to understand what they are doing differently and in what context (see Fig. 4.2).

Finland and the Netherlands account for the majority of the highest-rated components of national-level entrepreneurial ecosystem as shown in Table 4.2.

Due to the low scores for individual domains, the composite index for the national-level entrepreneurial ecosystem in Croatia is very low, as shown in Table 4.3.

[9] Croatia, Cyprus, Finland, France, Germany, Greece, Hungary, Ireland, Italy, Latvia, Lithuania, Luxembourg, the Netherlands, Poland, Romania, Slovak Republic, Slovenia, Spain, Sweden.

Table 4.2 Scores for entrepreneurial ecosystem on the national level, 2021

Items	EU countries with the highest score	Highest score in EU	Avg EU	Croatia
Entrepreneurial finance	Finland	7.13	5.04	4.30
Ease of access to entrepreneurial finance	Finland	6.61	4.70	3.73
Government policy: support, relevance	France	6.27	4.52	2.72
Government policy: taxes and bureaucracy	Netherlands	6.82	4.86	3.43
Government entrepreneurship programs	Netherlands	6.43	5.06	4.11
EE in pre-tertiary schools	Finland	6.09	3.35	2.71
EE in colleges and universities	Netherlands	6.07	4.62	3.41
Research and development transfer	Finland	5.97	4.49	3.29
Commercial and professional infrastructure	Finland	6.92	5.82	4.80
Ease of entry: market dynamics	Poland	6.35	4.79	5.83
Ease of entry: burdens and regulations	Netherlands	6.48	4.80	3.51
Physical infrastructure	Finland	8.59	6.70	6.00
Social and cultural norms	Netherlands	6.71	4.67	2.96

Source Singer et al. (2022)

Croatia had the lowest NECI score in 2019 (3.6) and in 2020 (3.7), which confirms that the national-level entrepreneurial ecosystem in which individuals perform entrepreneurial activities does not have supportive features.

Table 4.3 NECI (National Entrepreneurship Context Index) for EU countries, GEM 2021

Countries with the highest scores		Countries with the lowest scores	
The Netherlands	6.3	Croatia	3.9
Finland	6.2	Romania	4.0
Lithuania	6.1	Poland	4.2

0 = very inadequate insufficient status, 10 = very adequate sufficient status

Source GEM (Global Entrepreneurship Monitor) (2022). Global Entrepreneurship Monitor 2021/2022 Global Report: Opportunity Amid Disruption. London: GEM

4.3.3 Entrepreneurial Ecosystem at the Sub-National Level: Case of Osijek's EE Programs

The entrepreneurial ecosystem at the sub-national level is the immediate context in which individuals and institutions act, from looking for opportunities, to gauging their skills and fear of failure in the process of forming their intentions and gaining the necessary resources for starting a venture. Among Isenberg's (2010) principles for establishing an entrepreneurial ecosystem is to "shape the ecosystem around local conditions." Stam (2015, p. 2) emphasizes that the entrepreneurial ecosystem presents "a community of interdependent actors" in which entrepreneurial activities take place. The importance of local conditions and the community of interdependent actors confirm that entrepreneurial ecosystems are heavily dependent on the context. Stam and Spigel (2016) identified networks, leadership, finance, talent, knowledge, support services, formal institutions, culture, physical infrastructure, and demand as interdependent factors, whose design should provide productive entrepreneurship within a particular territory at the sub-national level.

Development of the entrepreneurial ecosystem at the sub-national level in Croatia began in the mid-1990s, when a group of researchers from the J.J. Strossmayer University of Osijek looked for ways to start the economic and social development process in eastern Croatia, where the university is located. The institutional infrastructure that they developed was the first entrepreneurial ecosystem at the sub-national level in Croatia:

- NOA, microcredit institution, 1996 www.noa.hr
- Center for Entrepreneurship, 1997 www.czposijek.hr
  - Franchise Centre, 2003
  - Family Business Forum, 2003
- Business Incubator, 2002 (content, procedures, organisational structure, governance structure) www.inkubator.hr
- University-based Master's Degree Program in Entrepreneurship, 2000 www.ices.hr

  - Undergraduate program, 2005
  - PhD in Entrepreneurship and Innovativeness, 2010

- CEPOR, SME and Entrepreneurship Policy Center, Zagreb, 2001 www.cepor.hr
- Strategic vision for the region, 2005–2015

  - Osijek-Baranja County: The county where young people want to live
  - Municipality of Osijek: From industrial to intelligent city

The timeline reflects the influence of the context where the immediate response was money. People needed money to start a business, to be self-employed. The first microfinance institution in Croatia was established based on the experience of Prof. Muhammad Yunus in running the micro-credit program in Bangladesh's Grameen Bank (1996), and with the USAID contribution of USD3 million. The first micro-loans were approved to those living in eastern Croatia, which was under UN administration, just prior to reintegration with Croatia.

Soon, it became obvious that without knowledge and skills, people did not know how to use money for entrepreneurship. The next move was to establish the first center for entrepreneurship in Croatia in 1997, based on the experience of the Small Business Development Center at the University of Wisconsin, US. The Open Society Institute (New York) provided financial support for training locals on the content and processes to set up such an entity. However, the initiators of those activities asked themselves whether adults needed training to acquire entrepreneurial skills, and what about the young people. That was how the first university-based EE program started in Croatia at the J.J. Strossmayer University of Osijek in 2000.

The lack of strategic vision for the region was a major impediment, so the immediate task was to strategize the participation of different stakeholders such as businesspeople, politicians, and researchers about the future of the Osijek-Baranja County and the City of Osijek. Young people were the most important stakeholders because the main issues were to make the county a place where young people would want to live in and to attract others to come, and how to move from an industrial to an "intelligent" city. Researchers looked at how some regions/cities in deprived places around the world managed to make changes, and this study was done in 2005–2015.

The next move was to establish the CEPOR—SME and Entrepreneurship Policy Center in 2001 as a think tank for researching policies for supporting entrepreneurship. Despite the wish of Osijek's researchers to set up CEPOR in Osijek, it was located in the capital, Zagreb.

The flow of events brought about the following observations:

- Most of the entrepreneurial activities arose out of necessity, which were at the same time opportunities for change.
- All activities were supported by the concept of effectuation (Sarasvathy, 2001), i.e., with almost no or minimal finances.
- Knowledge matters; money is not enough. The GEM surveys confirmed that people with higher education levels are also more entrepreneurially active (this will be shown in Sect. 4.4).

- Developing one's own entrepreneurial ecosystem involves supply and demand at the same time.
- An integrated approach is essential for developing the collaborative capacity of the entrepreneurial ecosystem at the sub-national level, i.e., the quadruple helix of academia, the business sector, government, and civil society. It is not enough to have institutions if they do not interact or collaborate; the cooperation of institutions is necessary to develop a supportive environment in which potential, new, and established entrepreneurs can operate.

All institutions developed as an immediate entrepreneurial ecosystem by the group of researchers from the J.J. Strossmayer University of Osijek, which still maintained their activities at a vibrant and sustainable level in 2022 when this survey was conducted. The cooperation between the university and these institutions has proven to be very productive. Many graduates of EE programs have taken on management positions or are in expert teams of various support institutions throughout Croatia, where such institutions were established using Osijek's experience.

4.4 EE and Entrepreneurship Activity: The Higher Educated Are More Entrepreneurially Active

The conceptual framework of the GEM and its data provide insights into the interconnectedness among many variables. The main findings relate to the following variables:

- Education matters: People with higher education express greater self-confidence, intentions to be entrepreneurially active, and perform entrepreneurial activities.
- Values matter: People with higher education are more appreciative of an entrepreneurial career and the societal status of being successful entrepreneurs.
- Entrepreneurship framework conditions matter.

Education influences how people express intentions toward venturing and how confident they are in their knowledge and skills (Figs. 4.4 and 4.5), but surprisingly it does not have much effect on how people view opportunities where they live (Fig. 4.3).

Perceived Opportunities Rate: Percentage of population aged 18–64 (excluding individuals involved in any stage of entrepreneurial activity) who see good opportunities to start a firm in the area where they live (Fig. 4.3).

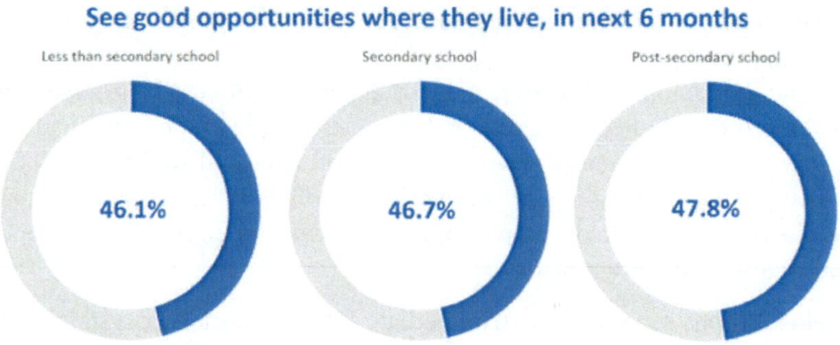

Fig. 4.3 People who view good opportunities where they live in the next six months (% of adult population). *Source* GEM database, Croatia

This data showing that the level of education does not have great impact on people perceiving good opportunities where they live could mean two things: there are no opportunities; or education fails to provide the knowledge and skills on how to recognize a good opportunity. It could be both, but it is an interesting challenge for further research.

The level of education plays a strong role in the level of self-confidence in one's abilities for starting a business, almost twice as much between those with less than secondary education and those with post-secondary education.

Perceived Capabilities Rate: Percentage of population aged 18–64 (excluding individuals involved in any stage of entrepreneurial activity) who believe they have the required skills and knowledge to start a business (Fig. 4.4).

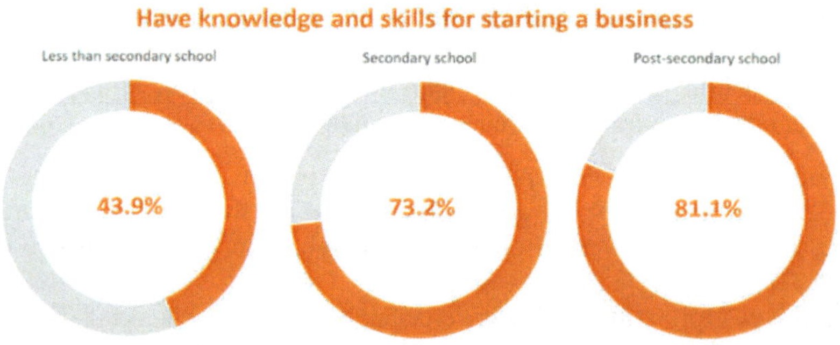

Fig. 4.4 Individuals with knowledge and skills for starting a business (% of adult population). *Source* GEM database, Croatia

The same applies to intentions of starting a business in the next three years, but at a significantly lower level than expressing self-confidence in one's abilities, which is understandable. One is less realistic when evaluating themselves, while more realistic when promising to act.

Entrepreneurial Intentions Rate: Percentage of population aged 18–64 (excluding individuals involved in any stage of entrepreneurial activity) who are latent entrepreneurs and who intend to start a business within three years (Fig. 4.5).

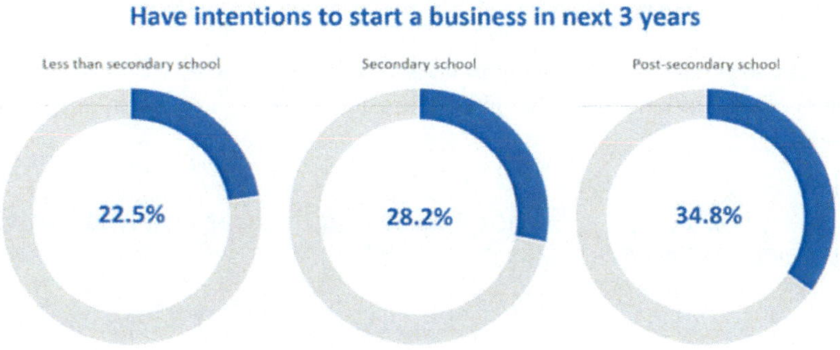

Fig. 4.5 Individuals with intentions to start a business in the next three years (% of adult population). *Source* GEM database, Croatia

Interestingly, the fear of failure is almost at the same level, regardless of the level of education. The intensity of the fear of failure in Croatia (49.9% of surveyed adults) is at the average level of EU countries (48.4%), but much higher than in the Netherlands (35.4%). The reason could be cultural, but education may also have neglected this aspect of entrepreneurial ability. This does not mean that people should be freed from the fear of failure, but that they should be able to understand the concept of calculated risks (Fig. 4.6).

Fear of Failure Rate: Percentage of population aged 18–64 (excluding individuals involved in any stage of entrepreneurial activity) who indicate that fear of failure would prevent them from setting up a business.

Fig. 4.6 Have fear of failure (% of adult population). *Source* GEM database, Croatia

Relationships between educational level and entrepreneurial activity are obvious in all three categories of entrepreneurial activity as defined by the GEM conceptual framework:

- New venture: total early-stage entrepreneurial activity (business venture up to 42 months of activity) (Fig. 4.7).
- Established venture: more than 42 months of activity (Fig. 4.8).
- Entrepreneurial employee activity: developing new products or new business units for the employer (Fig. 4.9).

Total Early-stage Entrepreneurial Activity (TEA) Rate: Percentage of population aged 18–64 who are either a nascent entrepreneur or owner-manager of a new business.

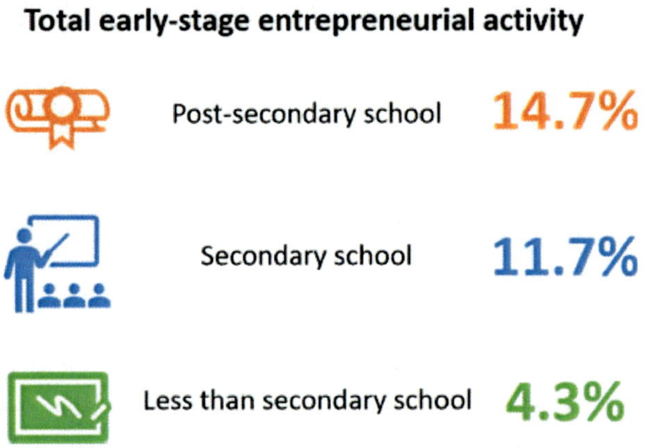

Fig. 4.7 Total early-stage entrepreneurial activity (% of adult population). *Source* GEM database, Croatia

Established Business Ownership Rate: Percentage of population aged 18–64 who are currently an owner-manager of an established business, i.e., owning and managing a business that has paid salaries, wages, or any other remuneration to the owners for more than 42 months.

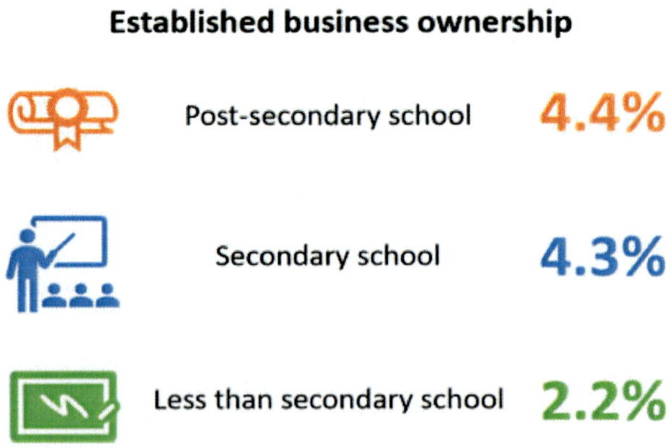

Fig. 4.8 Established business ownership (% of adult population). *Source* GEM database, Croatia

Entrepreneurial Employee Activity Rate: The rate of involvement of employees in entrepreneurial activities, such as developing or launching new goods or services, or setting up a new business unit, establishment, or subsidiary.

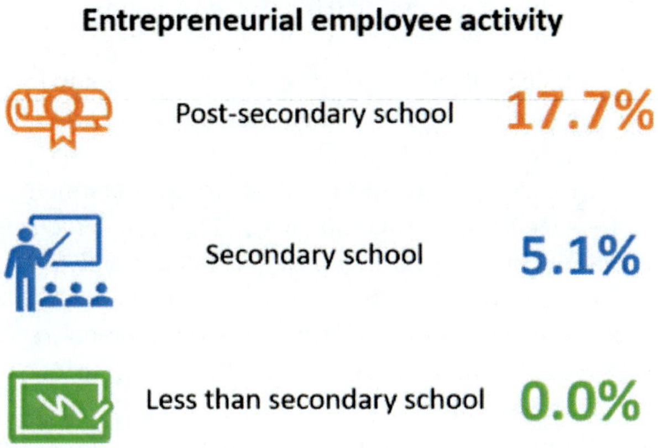

Fig. 4.9 Entrepreneurial employee activity (% of adult population). *Source* GEM database, Croatia

The strongest relationship between educational level and entrepreneurial activity is present in the category of entrepreneurial employee activity. Since Croatia is among the first five EU member states participating in the GEM survey for this form of entrepreneurial activity, it is particularly useful information for business owners/managers to introduce effective compensation schemes to motivate employees to make innovative contributions. This information should also be recognized by research institutions as well as policymakers to encourage collaboration between research institutions and businesses.

The analysis of the relationships between education and selected indicators of an individual's entrepreneurial profile (*attributes*: ability to recognize opportunities, possess intentions to start a venture, manageable fear of failure; *activity:* new venture, established venture, entrepreneurial employee activity) strongly confirms that education is important. At the same time, the analysis of the entrepreneurial ecosystem at the national level (using the GEM data on the quality of each domain of the ecosystem, Sect. 4.3.2) shows that college and university EE scored the lowest, indicating that this domain is a bottleneck impeding the supportive capacity of the entire ecosystem. This is a very important message to HEIs, and to the government, which is responsible for education policies.

4.5 The Challenge of the Relevance of EE at HEIs Arising from 2050 UN SDGs

The UN Sustainable Development Goals (SDGs) are the only global consensus on areas where everyone including individuals and institutions should contribute to and where collaborative actions are needed, in order to achieve people's and the planet's well-being. The identified targets are important milestones for checking on the progress in achieving them and for figuring out if new strategies are needed. Education (Goal 4: Quality Education) is one of the 17 goals in the 2030 Agenda for Sustainable Development, which was adopted by all UN member states in 2015. A long-term vision for 2050 is based on the same goals, but they are clustered according to their contribution to the well-being of people (Goal 4: Quality Education) and well-being of the planet (Fig. 4.10). Two goals relate to governance and means of implementation (Goal 17: Partnership, and Goal 16: Justice and Strong Institutions) with infrastructural roles.

In general, education and skills are identified as one of the six major drivers of societal transformations needed to ensure the achievement of 2050 expectations relating to the well-being of the people and that of the planet. In Croatia, Goal 4 (Quality of Education), according to the 2022 Sustainable Development Report (Sachs et al., 2022, p. 23), is assessed to be on track to achieve the target, but Goal 16 (Justice and Institutions) and Goal 17 (Partnership) are assessed as "significant challenges remain."

Based on the 2050 UN SDG, HEIs carry a great responsibility to be drivers of societal change across the globe. Tertiary education impacts almost everyone's

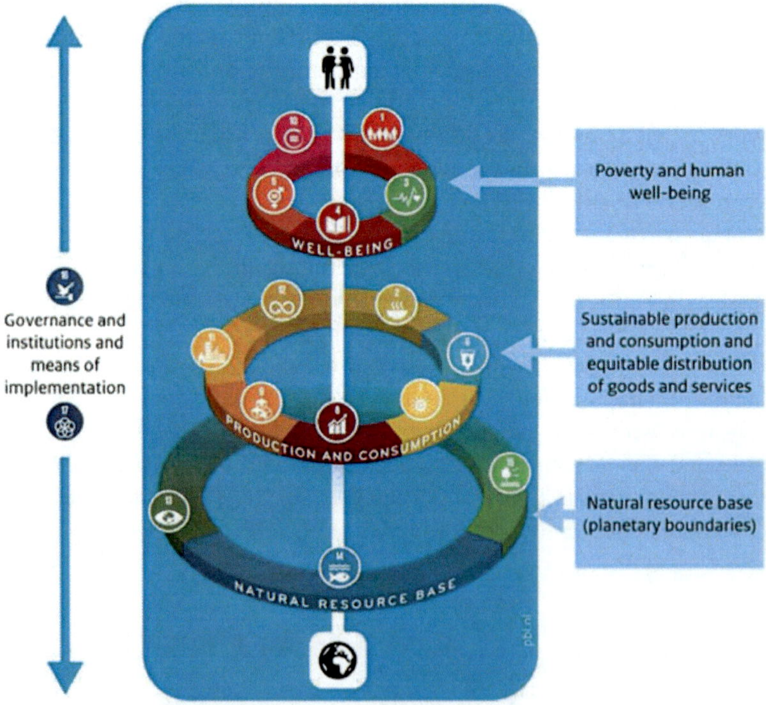

Fig. 4.10 2030–2050 UN sustainable development goals. *Source* PBL Netherlands Environmental Assessment Agency (2017) as presented in UN Environment (2019), Ch. 20 A Long-Term Vision for 2050, p. 474

lives, directly or indirectly, therefore it is important that HEIs provide access to relevant knowledge. According to the OECD (2018, p. 5), in order to be prepared for the future, "individuals have to learn to think and act in a more integrated way, taking into account the interconnections and inter-relations between contradictory or incompatible ideas, logics and positions, from both short- and long-term perspectives. In other words, they have to learn to be systems thinkers." The concept of transformative competences, introduced by the OECD (2018, p. 5), emerged from the growing need for young people to be able to create new value, reconcile tensions and dilemmas, and take responsibility. These are qualities almost identical to entrepreneurial competences: proactiveness, innovativeness, and being responsible in pursuing opportunities, regardless of resources.

Acting in a more integrated way, or to be systems thinker, is a fundamental concept of EE, as presented through analyses of the internal structure of EE programs and the entrepreneurial learning ecosystem at the national and sub-national levels. Everything is interconnected and requires a departure from the traditional definition of curriculum based on courses and disciplines. "Disciplinary knowledge will continue to be important, as the raw material from which new knowledge is developed" (OECD, 2018, p. 5), but the process of producing relevant knowledge should become the norm in EE and the core content of the new "social contract" between HEIs and students. This process is in line with SDG 4, and its target 4.4: "By 2030, substantially increase the number of youth and adults who have relevant skills, including technical and vocational skills, for employment, decent jobs and entrepreneurship."

UNESCO Chairs in Entrepreneurship Education could lead the way in promoting transformation for higher education and, through international networking, in building capacity for a new social contract. This would be based on the following:

- Implement Schumpeter's concept of *creative destruction* to open new ways of organizing knowledge and pedagogy on campus within the framework of the entrepreneurial ecosystem, based on the quintuple helix of academia, business, government, civil society, and planet.
- Adopt Stevenson's approach that entrepreneurship is *pursuing opportunities* regardless of resource limitations, backed by Sarasvathy's concept of *effectuation.*
- Perceive entrepreneurship as a true *mobilizer* and *equalizer* of opportunities in society (Timmons, 2008).
- Make the universities' mission focus on competences for *identifying and solving problems*, and *exchanging ideas* (Reich, 1992).

A Never-ending Story

The legacy of J.J. Strossmayer University of Osijek: Pioneering work in developing research-based EE in Croatia.

This venture was triggered by the 1991–1995 war in eastern Croatia where the university is located, but EE was implemented and sustained by a group of local researchers with a strong international network, which became important sources of learning.

***Lessons learnt*:**
Creative destruction, networking, learning from others, contextualization, vision, not giving up, courage, and kindness.

What is to come:

Provide access to EE to *all* students across campus through the International Centre for Entrepreneurial Studies.

Strengthen the international dimension of the programs through virtual learning and using English as a working language.

Intensify training of educators in the production of relevant knowledge by using team teaching, case studies, and drama as learning techniques, among other new learning methods.

Redesign undergraduate and graduate EE programs by including a strong component of working on local issues relating to UN SDGs.

Keep giraffes as mentors, forever:

The doctoral program, Entrepreneurship and Innovativeness, adopts giraffes as its brand because of its features that are very relevant to EE:

- The giraffe is the tallest terrestrial animal, so it has the best overview of opportunities and dangers.
- The giraffe usually sleeps with one eye open and its ears remain alert: always be alert to changes.
- The giraffe's height is a disadvantage when it comes to drinking water: they have to spread their front legs and are then vulnerable to predators. Often, one giraffe stands to keeps watch while the others drink—a show of teamwork.
- The giraffe has a relatively small heart, but with a very strong beat: a strong heart, or passion, is important for any venture. And that was the case when EE programs were launched in a remote, war-torn region without money, but with great passion.

References

Act on Scientific Activity and Higher Education (Official Gazette, 123/2003).

Agency for Science and Education—Statistics. (2022). https://www.azvo.hr/en/higher-education/statistics

Croatian Bureau of Statistics (Enumerated persons, households and housing units, 2021. Census, first results, by settlements).

European Union. (2021). *European innovation scoreboard 2021*. Publication Office of the European Union.

European Commission, Directorate-General for Internal Market, Industry, Entrepreneurship and SMEs. (2021). *Regional innovation scoreboard 2021*. https://doi.org/10.2873/674111

European Commission, Directorate-General for Education, Youth, Sport and Culture, Halbfas, B., Volkmann, C., Bischoff, K., et al. (2015). *Supporting the entrepreneurial potential of higher education: Final report: Executive summary, version 1.1*.https://doi.org/10.2766/508157

European Commission, Directorate-General for Education, Youth, Sport and Culture. (2015). *Supporting the entrepreneurial potential of higher education: Final report appendix: Case studies*, University of Osijek, Croatia: Developing entrepreneurship education from scratch over time, pp. 364–390 Publications Office, https://doi.org/10.2766/441172

GEM (Global Entrepreneurship Monitor). (2022). *Global Entrepreneurship Monitor 2021/2022 global report: Opportunity amid disruption*. GEM.

Gibb, A. A. (1993a). Enterprise culture and education: Understanding enterprise education and its links with small business, entrepreneurship and wider educational goals. *International Small Business Journal*, 11(3): 11–34.

Gibb, A. A. (1993b). Small business development in Central and Eastern Europe: Opportunity for a rethink? *Journal of Business Venturing, 8*(6), 461–486.

Gibb, A. A. (2000a). SME policy, academic research and the growth of ignorance, mythical concepts, myths, assumptions, rituals and confusions. *International Small Business Journal, 18*(3), 13–25.

Gibb, A. A. (editor and main contributor). (2000b). Entrepreneurial training for the growth of small and medium-sized enterprises. Lessons from Central and Eastern Europe. European Training Foundation Turin.

Gibb, A. A. (2002). Creating conducive environments for learning and entrepreneurship: Living with, dealing with, creating and enjoying uncertainty and complexity. *Industry and Higher Education, 16*(3), 135–148.

Gibb, A. A. (2005) . *Towards the entrepreneurial university: Entrepreneurship education as a lever for change* (National Council for Graduate Entrepreneurship, policy paper 003).

Gibb, A. A., & G. Haskins. (2013). The university of the future—An entrepreneurial stakeholder learning organisation? In A. Fayolle & D. T. Redford (Eds). *Handbook of research in entrepreneurial education Volume 4—Entrepreneurial university handbook*. Edward Elgar.

Government of Croatia. (2010). *Strategija učenja za poduzetništvo 2010–2014* (Strategy on learning for entrepreneurship 2010–2014).

Government of Croatia. (2014). *Strategy for education, science and technology—New colours of knowledge* (Strategija obrazovanja, znanosti i tehnologije - Nove boje znanja), Official Gazette, 124/2014.

Government of Croatia. (2021a). *Nacionalna razvojna strategija Republike Hrvatske do. 2030.* godine, NN 13/2021 (National development strategy of the Republic of Croatia until 2030, Official Gazette 13/2021).

Government of Croatia. (2021b). *Nacionalni plan oporavka I otpornosti 2021–2026* (Recovery and resilience plan for Croatia 2021–2026).

Harmeling, S. S., Oberman, S., Venkataraman, S., & Stevenson, H. H. (2004). *That my neighbor's cow might live: Effectuation, entrepreneurship education and economic development in Croatia.* Babson College—Kauffman Foundation Entrepreneurship Research Conference, Glasgow, UK, June 2004. Frontiers of Entrepreneurship Research 2004, conference proceedings (ISBN: 0-910897-25-5).

HBS Alumni Bulletin. (December 1, 1996). *Entrepreneurship at HBS*. https://www.alumni.hbs.edu/stories/Pages/story-bulletin.aspx?num=6150

Hegel, G. W. F. (2000). *Phenomenology of spirit* (T. Pinkard, Trans.). Cambridge University Press.

Isenberg, D. (2011, May 12). *The entrepreneurship ecosystem strategy as a new paradigm for economic policy: Principles for cultivating entrepreneurship* (pp. 1–13). Institute of International and European Affairs, Dublin, Ireland.

Isenberg, D. J. (2010). The big idea: How to start an entrepreneurial revolution. *Harvard Business Review, 88*(6), 41–50.

Ministry of Entrepreneurship and Crafts. (2013). *Strategija razvoja poduzetništva u. Republici Hrvatskoj 2013–2020* (Strategy for developing the entrepreneurship in the Republic of Croatia 2013–2020, Official Gazette 136/2013).

Ministry of Science and Education. (2018). *Strateški plan 2018–2020* (Strategic plan 2018–2020).

Ministry of Science and Education. (2019). *Strateški plan 2020–2022* (Strategic plan 2018–2020).

OECD (Organisation for Economic Cooperation and Development), European Commission, European Training Foundation and European Bank for Reconstruction and Development. (2009). *SME Policy Index 2009: Progress in the implementation of the European charter for small enterprises in the Western Balkans*. OECD.

OECD. (2018). *The future of education and skills—Education 2030*. OECD.

OECD/European Union. (2019). *Supporting entrepreneurship and innovation in higher education Croatia*. https://heinnovate.eu/en/heinnovate-resources/resources/oecdec-supporting-entrepreneurship-and-innovation-higher-education-4

Pfeifer, S., Singer, S., Šarlija, N., & Oberman Peterka, S. (2021). Perception of the national entrepreneurship conditions—Differences across time and expert specialization. *Sciendo, South East European Journal of Economics and Business, 16*(1), 1–17.

Reich, R. B. (1992). *The work of nations: Preparing ourselves for 21st century*. Vintage.

Rinaldi, S., Klenha, V., Feiler, L., & Petkova, E. (2012). *Croatia—Review of human resources development*. European Training Foundation.

Research projects financed by the Croatian Ministry of Science, on entrepreneurship, from 1992. The Constitution of the Republic of Croatia (1990, 1998).

- Theoretical and institutional foundation of entrepreneurial economy, 1991–1996 (Teorijske i institucionalne pretpostavke poduzetničke ekonomije), principal researcher S. Singer, 13 researchers.
- Paradigm and praxis of entrepreneurial economy, 1996–2002, 2002–2006 (Paradigma i praksa poduzetničke ekonomije, principal researcher S. Singer, 13 researchers.
- Entrepreneurship – mobiliser of social integration, 2006–2012 (Poduzetništvo – mobilizator socijalne integracije), principal researcher S. Singer, 13 researchers.
- Development and application of growth potential prediction models for small and medium enterprises in Croatia, 2014–2018, principal researcher N. Šarlija, 8 researchers.

Sachs, J., Lafortune, G., Kroll, C., Fuller, G., & Woelm, F. (2022). *Sustainable Development Report 2022. From crisis to sustainable development: The SDGs as roadmap to 2030 and beyond*. Cambridge University Press.

Sarasvathy, S. D. (2001). Causation and effectuation: Toward a theoretical shift from economic inevitability to entrepreneurial contingency. *The Academy of Management Review, 26*(2), 243–263.

Schumpeter, J. A. (1934). *The theory of economic development*. Harvard University Press.

Shane, S., & Venkataraman, S. (2000). The promise of entrepreneurship as a field of research. *Academy of Management Review, 25*(1), 217–226.

Singer, S., et al. (1987–1990). *Zakon vrijednosti u funkciji upravljanja razvojem (Law of value in managing development)* financed by the Ministry of Science of the Republic of Croatia, as part of the fundamental research in economics.

Singer, S., Šarlija, N., Pfeifer, S., & Oberman Peterka, S. (2022). *ŠtočiniHrvatsku. (ne)poduzetničkomzemljom? GEM Hrvatska 2021*. Cepor.

Skjolstrup, K. (2008, May). *Human resource development country analysis for Croatia* (ETF Working Paper) .

Stam, E. (2015). Entrepreneurial ecosystems and regional policy: A sympathetic critique. *European Planning Studies, 23*(9), 1759–1769.

Stam, F. C., & Spigel, B. (2016). *Entrepreneurial ecosystems* (USE Discussion Paper Series, 16), pp. 1–18.

Sveučilište Josipa Jurja Strossmayera u Osijeku. (2021). *Strategija 2021–2030* (Strategy 2021–2030). http://www.unios.hr/wp-content/uploads/2022/02/strategija-sveucilista_20211124.pdf.

Sveučilište Josipa Jurja Strossmayera u Osijeku. (2020). Godišnjeizvješćerektorasveučilišta. o raduiposlovanjuSveučilišta Josipa Jurja Strossmayera u Osijeku za akademsku 2018./2019 godinu (Annual rector's report on activities of the J.J. Strossmayer University in Osijek for academic 2018/2019 year).

Timmons, J. A. (2008). *New venture creation: Entrepreneurship for the 21st century*. McGraw-Hill.

Vojnić, D. (1993). Economics and politics of transition—The road to the welfare state and economy. In Ž Puhovski, I. Prpić, & D. Vojnić (Eds.), *Politics and economics of transition* (pp. 245–373). Informator.

Chapter 5
Entrepreneurship Education in Finland

Jarna Heinonen and Sanna Ilonen

This chapter presents the state-of-the-art entrepreneurship education (EE) in Finland through lenses applied in this volume. We first provide background by describing the context for EE in Finland. We continue by discussing the internal development of EE in the individual HEIs based on the information provided in their websites in Fall 2021.Then, we describe the supporting systems of EE in HEIs and the characteristics of EE by highlighting an example of an entrepreneurial university. We conclude with alternative futures toward higher quality of EE.

5.1 Context

After the Second World War, Finland developed from a poor agricultural country to a forerunner in high technology and well-being. Education played a role in improving the cultural level of the country. Finland's education system is publicly financed, including universities. The latter currently rely heavily on public funds, although some of them were originally established by private initiatives and funds.

Economic education in HEIs began in business schools. During the 1950s–1960s, the focus was on educating economic issues, whereas in the 1980s the emphasis was on providing EE. Finally, from the 1990s onward, the concept of enterprise education evolved in parallel with EE with slightly different meanings and foci (Ministry of Education & Culture, 2009).

J. Heinonen (✉) · S. Ilonen
University of Turku, School of Economics, Turku, Finland
e-mail: jarna.heinonen@utu.fi

S. Ilonen
e-mail: sanna.ilonen@utu.fi

X. Xu, *Comparative Entrepreneurship Education*,
https://doi.org/10.1007/978-981-99-1835-5_5

EE, or rather the first mentions of entrepreneurship in the national education program, took place in 1985. At the national level, the notion of EE to support individuals for earning a living independently or to improve employability was launched in the 1990s as a remedy to the economic and financial crisis in Finland. Entrepreneurship was meant to provide citizens with self-employment and entrepreneurial opportunities, as a record number of individuals was laid off (Ministry of Education & Culture, 2009). Furthermore, the idea of intrapreneurship was apparent as measures also addressed individual employability in the national education program in 1994.

In 1999, the Ministry of Trade and Industry launched the two-year Entrepreneurship Programme—a cross-sectional and horizontal approach to entrepreneurship—with the aim of increasing economic growth, employment, and diversifying the industrial structure. The focus was on SMEs: to remove barriers and reduce administrative burden in order to incentivize small firms to grow. As for EE, the program focused on developing the education system to support entrepreneurship and the administration for this end, not on encouraging individuals to undertake entrepreneurship per se. The Entrepreneurship Programme was a precursor of specific entrepreneurship policy initiatives included in official government programs from the early 2000s (Heinonen & Hytti, 2016). Through the Entrepreneurship Programme, the idea of cross-sectoral and horizontal collaboration between ministries and with stakeholders was seeded into Finnish entrepreneurship policy mission and its implementation. Given this, EE policies were also guided and implemented as a part of ministerial collaboration. In the following, we first give an overview of the government-level entrepreneurship policies based on government programs to understand the main problems and challenges to be addressed by policies,[1] and then examine specific EE policies.

An important step was taken in 2003 when an entrepreneurship program was launched as a part of the government program. The aim of the program was to secure economic growth, employment, and investments in Finland. The focus was on individuals, particularly potential entrepreneurs, rather than firms. The program provided incentives and education for an entrepreneurial career. This was done, for example, by introducing guidelines for enterprise education at different educational levels.

In 2007, the new government introduced a policy program on work, entrepreneurship, and work-life. The aim was to promote economic growth, employment, and a welfare society. The policy was broadened to include work and the working lives of individuals and was less focused on entrepreneurship. To improve the employment situation, it emphasized the responsibility of entrepreneurs and small businesses to create more and better jobs, and the responsibility of individuals for remaining employable. The latter can be understood as a way to encourage individuals to take on entrepreneurship, or rather intrapreneurship, as it views every individual as an entrepreneur in charge of one's own employment. The focus of the program was on

[1] The developments were derived from the study of Heinonen, J., & Hytti, U. (2016). Entrepreneurship mission and content in Finnish policy programmes. *Journal of Small Business and Enterprise Development, 23*(1), 149–162.

employees and entrepreneurs. Flexicurity, i.e., balancing flexibility and security at work, helped remove barriers from employment and improve individual employability. Furthermore, the desirability of entrepreneurship was enhanced through EE, the offerings of which were broadened in addition to intensifying collaboration between education and businesses, for example.

In the aftermath of the 2011 global financial crisis, the government again assumed the role of fighting poverty and social exclusion as well as balancing public finance and employment. The government program did not have a dedicated program for entrepreneurship, but the key project on the enhancement of sustainable economic growth, employment, and competitiveness was launched with the aim to secure socially and ecologically sustainable economic growth, employment, and development of new industries. This focus was a clear response to the global economic crises after a long period of continuous growth in the 2000s. Entrepreneurship was again expected to create more and better jobs to tackle the challenges of an ageing population, modest economic growth, and social exclusion. Entrepreneurship was thus presented as a remedy rather than an opportunity for society.

Similarly, in 2015, no dedicated entrepreneurship program was included in the government program. However, the entrepreneurship package introduced eight wider measures to help existing entrepreneurs and businesses. Measures such as the renewal of growth services for businesses and lowering the threshold for employing the first employee targeted at existing businesses with recruiting, growth, and internationalization potential. EE was not emphasized in the government's entrepreneurship package.

In 2019, entrepreneurship strategy was included in the government program to increase trust among companies that it was worth creating employment, developing businesses, and investing as well as becoming an entrepreneur in Finland. The eight strategic themes of the entrepreneurship strategy were again targeted at existing and potential entrepreneurs. EE was not explicitly in the agenda of the entrepreneurship strategy headed by the Ministry of Economic Affairs and Employment, although many other ministries, including the Ministry of Education and Culture, were involved in its planning and execution.

The emphasis of Finland's entrepreneurship policy has evolved over the years under different governments and political parties. Although entrepreneurship policies have addressed a variety of issues during the years, the primary focus began with administrative support for SMEs. Then, first, it moved to emphasizing entrepreneurial careers and individuals; second, working life; third, it considered entrepreneurship as a remedy to the financial and economic crisis; and finally it focused on existing and potential entrepreneurs and growth-oriented businesses. The shifts in focus reflect the economic situation of Finland rather than political power relationships. Although the focus on EE was limited in the above entrepreneurship policies, particularly in the recent ones, EE has been more visible in the national guidelines of the Ministry of Education (and Culture).

Through EE, the Ministry of Education has supported entrepreneurship and innovation in Finland. From the beginning, the aim has been to make entrepreneurship more attractive as a career option and to enhance entrepreneurial attitudes and competences throughout the education system. The emphasis varies at different educational levels. In general, the focus in primary education is on enhancing entrepreneurial attitudes, basic knowledge and skills as well as an entrepreneurial mode of operation. In secondary and higher education, the focus is on the development of knowledge and skills, including entrepreneurial competences.

For HEIs, EE focuses on strengthening entrepreneurial attitudes among students and teachers; boosting innovation and business ideas; promoting the utilization and commercialization of research findings and academic expertise; academic entrepreneurship; and promoting collaboration with businesses, science parks, and technology centers. For universities, in particular, these imply the exchange of knowledge between academia and businesses as well as strengthening the entrepreneurial competences of researchers (Ministry of Education, 2004; Ministry of Education & Culture, 2009). These national guidelines have been adopted from EU policies and guidelines, and further accommodated for regional and local level strategies and execution.

The most recent national guidelines (Ministry of Education & Culture, 2017) approach EE from various perspectives:

1. *Strategic level and leadership*: Defines the basics through planning, resource allocation, and evaluation of activities. Furthermore, issues relating to staff competences and collaboration across fields and with working life are highlighted.
2. *Training for education and teaching staff*: Defines the access of teachers to EE training at all educational levels.
3. *Training that supports entrepreneurship*: Defines the access of EE at all levels of education and related collaborations.
4. *Learning environments*: Focuses on a culture of experimentation and versatile and innovative learning environments and situations.

These EE guidelines of the Ministry of Education and Culture guide and develop measures promoting entrepreneurship and enterprise education at different educational levels. Furthermore, they serve as a part of the ministry's information guidance and demonstrate public commitment to introduce EE in a variety of ways, from kindergarten to higher education. Next, we will look at higher education and universities to understand how EE has developed there.

5.2 Internal Development of EE in HEIs

Finland has a strong strategic emphasis for EE at all educational levels, including higher education (Ministry of Education, 2004; Ministry of Education & Culture, 2009, 2017). Due to long-term efforts, Finnish entrepreneurship culture has improved significantly in recent years (OECD, 2021). Finland's higher education field consists of two types of institutions: universities and universities of applied sciences. Universities focus on scientific research and education, while universities of applied sciences offer more pragmatic education that responds to the current needs of working life. Finland has 13 scientific universities spread across the country. The Ministry of Education and Culture has actively monitored the development of EE in the higher education system, and two reports produced in 2005 and 2016 give overviews of EE in universities (Ministry of Education & Culture, 2005, 2016). In the following, we analyze the internal development of EE in the Finnish universities based on the outcomes of these two policy reports. Then, we describe the current state of EE in Finnish universities. The descriptions of the current state is handpicked from each university's website in Fall 2021. The section ends with a synthesis of the policy developments and the current state.

5.2.1 EE in Finnish Universities in the Early 2000s

In 2005, the Finnish university field consisted of 21 universities located all over Finland. The report emphasizes that there seemed to be an extensive shift from salary work toward "an entrepreneurial society," but universities were still not well prepared nor equipped to facilitate this shift. At the same time, however, many distinct EE-related expectations were directed toward the Finnish university field. Even then, EE was seen as a "relatively marginal phenomenon in Finnish universities." Nonetheless, many universities considered entrepreneurship to be an important issue that was closely related to the university's third mission of social interaction. No specific entrepreneurship strategies were identified in the universities. Moreover, there were no explicit objectives for EE. In terms of entrepreneurship, universities' role was seen as a creator of new scientific knowledge in entrepreneurship; a compiler and distributor of entrepreneurial knowledge for the purposes of the business sector; being an EE teacher; or solver of local needs. The policy report raised the importance of multidisciplinarity in promoting EE. The integration of EE with non-business subjects was mentioned as one of the most important tools to facilitate EE in universities. In addition, there seemed to be a gap between academia and business reality. This gap was recommended to be solved with tighter interaction and innovative solutions. The report also highlighted the importance of EE in both universities and universities of applied sciences. The report stated that the key challenges of EE concerned financing of related programs, facilitation of interaction between universities and stakeholders, and reforming ongoing curriculum to reflect the importance of EE in universities.

In 2016, the Finnish university field consisted of 14 universities. That year's report highlighted that EE had been widely offered in Finnish universities at least in individual entrepreneurship courses, which implied there had been tremendous development between 2005 and 2016. The aim of EE was particularly to support the formation of new businesses and startups. Finnish universities had noticed the importance of pedagogy in facilitating entrepreneurial learning. However, there seemed to be a lack of such pedagogical training for entrepreneurship educators. Universities were also lagging behind universities of applied sciences in terms of collaboration with the business sector. However, universities and universities of applied sciences had some collaboration in terms of EE, but it was suggested that such collaboration could be further developed in the future. Multidisciplinary entrepreneurship research with diverse foci was conducted in many universities, which formed a good stepping stone to developing entrepreneurship-related education. In addition, universities were interested in the commercialization of research and had invested in different types of innovation and patenting services, and support structures. Student-led entrepreneurship societies had boomed all around Finland, and each town with a university campus had its own student-led entrepreneurship society. In 2005, none of these student-led entrepreneurship societies had existed. Universities had some connections and collaborations with the entrepreneurship societies, but it seemed that the collaboration was not yet that systematic, particularly outside capital region, Helsinki. It was suggested that the assessment and evaluation of the impact of EE would require further attention as universities applied very different measures in analyzing the impact. Some focused on the number of students in EE courses, while some looked at the employability and the number of businesses and startups after graduation.

5.2.2 *Current State of EE in Finnish Universities*

The following provides a brief description of all 13 Finnish universities and their EE activities. The information is not exhaustive, but focuses on core EE activities published in the universities' webpages in August/September 2021. The data, thus, reflect the ways in which the universities presented their EE activities and hence the articulation of the role of EE. The data was not complemented or verified with interviews, for example.

5.2.2.1 Aalto University

Aalto University is composed of six schools with close to 17,500 students and 4,000 faculty members. It is Finland's second-largest university. Its main campus is located at Otaniemi, Espoo. Aalto University is one of the best-known universities in Finland in terms of its entrepreneurship-related activities. Every student in Aalto can complete a minor or take individual courses in entrepreneurship. These are organized by the Aalto Ventures Programme, founded in 2012 as a joint initiative of students and faculty across the university. In addition, Aalto offers another entrepreneurship related minor: Aaltonaut, a bachelor's minor program in Interdisciplinary Product Development, launched in 2013. Aalto has a master's program in Entrepreneurship and Innovation Management, but this is planned to be discontinued. Students can graduate from the MSc program in Entrepreneurship and Innovation Management until July 31, 2023. Startup-oriented researchers can receive help in commercializing their research at the Aalto Startup Center, which is a hybrid accelerator offering its own incubator and accelerator services as well as several partnership programs. It focuses on sustainable research-based and innovative deep-tech startups. Aalto's strength lies in its student-led entrepreneurship society known as Aaltoes, the first student-led entrepreneurship society in Finland. Aaltoes is active in organizing entrepreneurship-related events and boosting entrepreneurship within Aalto University. Aaltoes played an important role in the development of the awarded, international startup event Slush. In addition, it runs KIUAS, which is an incubator, accelerator, and bootcamp. It also organizes FallUp, Europe's biggest student-run entrepreneurship event for students, as well as hackathons such as Junction and Dash. Aalto University provides opportunities for internship in startups in Asia and Silicon Valley in a program called Startuplifers. The program was established from the initiative of students and it is now open for students in several universities in Finland. Aalto University also contributes to EE in the country by offering a free startup online course for anyone interested. It is an introductory course to startup entrepreneurship where one can learn the basics of setting up a small business. Finally, Aalto University has active research groups that study entrepreneurship.

5.2.2.2 University of Helsinki

The University of Helsinki is the largest university in Finland with close to 32,000 students and 8,000 faculty members. It has 11 faculties and teaching in our campuses in Helsinki. The university provides individual courses in EE for all interested students. The Faculty of Agriculture and Forestry offers an entrepreneurship-related module consisting of several entrepreneurship courses. The faculty also plans to develop a hub where companies and students can better interact and discuss projects and real-life cases. The Ruralia Institute provides entrepreneurship-related courses from the perspective of rural studies. It coordinates the Co-op Network Studies, which offers courses and modules relating to the co-operative sector and social economy. Other participants in the network are the Aalto University School of Business, University of Jyväskylä, LUT University, University of Oulu, and University of Tampere. The Helsinki Think Company plays an important role in supporting and organizing courses and programs in collaboration with the University of Helsinki to encourage students to become entrepreneurs and supporting startup teams at different stages of development.

5.2.2.3 University of Eastern Finland

The University of Eastern Finland has approximately 16,000 students and close to 3,000 faculty members. It is composed of four faculties and its campuses are located in two cities: Joensuu and Kuopio. The university provides minor and individual courses in entrepreneurship for all students. Organized by the business school, the minor in entrepreneurship deepens students' understanding of the basics and requirements of generating, managing, and developing businesses. There are also initiatives relating to entrepreneurship studies for specific programs—for instance, in bioeconomy. Major degree studies in entrepreneurship used to be offered, but the program was discontinued during the structural development of the university field in 2010. The Joensuu campus collaborates with SPARK Joensuu, which offers entrepreneurial studies and startup business activities in science park facilities. SPARK is open to students from different levels of education. The University of Eastern Finland has two student-led entrepreneurship societies: one in Kuopio (KuopioES) and another in Joensuu (Joensuu Entrepreneurship Society). The university's business school is active in entrepreneurship-related research.

5.2.2.4 University of Jyväskylä

The University of Jyväskylä has approximately 14,000 students and six faculties with around 2,600 faculty members. Entrepreneurship is highlighted in the university's strategy and implemented as a cross-sectional feature for all its educational programs. The university provides minor and individual courses for anyone interested in entrepreneurship. It also offers a master's program in International Business and Entrepreneurship. The university collaborates closely with other educational institutes in terms of entrepreneurship studies. For instance, Edufutura—a joint initiative of the university and the Jyväskylä Educational Consortium Gradia, a university of applied sciences—provides general upper secondary education and vocational education, and organizes courses for university students. Students can participate in the Startup Factory, an incubator where initial-stage business ideas can be developed into companies in cooperation with educational institutes. Jyväskylä Entrepreneurship Society is a student-led entrepreneurship society in the region. It organizes activities that may be integrated with entrepreneurship studies. The University of Jyväskylä, particularly its business school, is active in entrepreneurship-related research.

5.2.2.5 University of Lapland

The University of Lapland is located in Rovaniemi. It is a relatively small university, consisting of four faculties with approximately 4,300 students and 450 faculty members. It is the northernmost university in the European Union. The university offers minor and/or individual courses in entrepreneurship. Entrepreneurship studies, which account for 60 ECTs, are organized by the open university. The university has provided EE in teacher education for decades. The student-led entrepreneurship society LaplandES is one of the newest entrepreneurship societies in Finland. It has organized hackathons and other entrepreneurship-related events for students. The University of Lapland has active entrepreneurship research focused on EE.

5.2.2.6 Lappeenranta University of Technology

The Lappeenranta University of Technology has approximately 5,300 students and almost 1,000 faculty members across three schools. The campuses are located in the cities of Lappeenranta and Lahti. The university has a strong strategic emphasis in entrepreneurship in terms of sustainability and renewal of industries. It offers entrepreneurship studies and a minor in entrepreneurship to all students. Moreover, it has two master's programs focused on entrepreneurship: International Business and Entrepreneurship, and an Entrepreneurship major in Engineering. The on-campus J. Hyneman Center brings together students, university research and skills as well as companies. It aims to support student entrepreneurship and university innovation. The student-led LUT Entrepreneurship Society, LUTES, organizes entrepreneurship-related events and programs for entrepreneurship-oriented students. The university has active entrepreneurship research. For instance, they have been coordinating the national data collection known as the Global University Entrepreneurial Spirit Students' Survey for years and have been actively developing a measurement tool for EE.

5.2.2.7 University of Oulu

The University of Oulu has close to 14,000 students and around 3,400 faculty members across eight faculties. It offers minor and individual courses in entrepreneurship for all students. The minor was piloted in 2016 (Ministry of Education & Culture, 2016). The student-led Oulu Entrepreneurship Society organizes entrepreneurship events and training such as Human Accelerator, Startup Weekend Oulu, and Idea Accelerator. Students can also take part in Demola, an open innovation platform. The university has been involved with the Polar Bear Pitching event, which has become fairly popular nationally and globally in recent years. The University of Oulu has active entrepreneurship-related research in its business school.

5.2.2.8 Hanken School of Economics

Hanken is a small Swedish-speaking university with approximately 2,500 students and around 250 faculty members, operating in both Helsinki and Vaasa. It offers a major degree in entrepreneurship and management. Its student-led Hanken Entrepreneurship Society was founded in 2010. Hanken has an incubator called Hanken Business Lab, which is a new form of incubator that aims to help startups, scale-ups, nonprofit organizations, and individuals achieve significant growth at Hanken. The Hanken Business Lab has operations in both university campuses. Hanken has active research in entrepreneurship.

5.2.2.9 University of Arts

The University of Arts is a small university based in Helsinki. It has fewer than 2,000 students and around 750 faculty members across three academies. It offers a master's degree program in Arts Management, Society and Creative Entrepreneurship.

5.2.2.10 Tampere University

Tampere University is a community of 21,000 students and close to 4,000 faculty members in seven faculties. The university was established in 2019 through a merger of the University of Tampere and the Tampere University of Technology. The university has a strategy for promoting sustainable entrepreneurship. It provides minor and individual courses in sustainable entrepreneurship for any students who are interested. The university collaborates with HUBS, which promotes sustainable entrepreneurship and offers support and optional studies in entrepreneurship and innovation to all students of Tampere University and the Tampere University of Applied Sciences. Furthermore, the student-led Tampere Entrepreneurship Society organizes entrepreneurship-related events and programs for entrepreneurship-oriented students. Students can also take part in Demola, an open innovation platform.

5.2.2.11 University of Turku

The University of Turku is a multidisciplinary university with approximately 20,000 students and around 3,400 faculty members across eight faculties. It is an entrepreneurial university with a strategic aim to facilitate entrepreneurship awareness and enhance EE in its all faculties. The purpose of EE in the university is to boost entrepreneurial behavior across society by generating new ventures and supporting established business activities as well as by enhancing entrepreneurial culture and attitudes among students and faculty members. All the faculties offer study programs that include courses with an entrepreneurship theme or an entrepreneurship-related focus. The business faculty offers individual courses on entrepreneurship as well as entrepreneurship as a major and minor. Moreover, students across faculties can take entrepreneurship as a minor, or some individual entrepreneurship courses. All business students have an obligatory course in entrepreneurship. In the faculty of science and engineering there is also a minor study module, Innovation and Business Creation, included in seven master's degree programs. In addition, the business school offers faculty-specific entrepreneurship studies—in medicine, for example. Students can participate in the student-led entrepreneurship society Boost Turku, which aims to inspire and help young students from different HEIs to become startup entrepreneurs by organizing events and programs. Boost Turku collaborates closely with the university and students can acquire study points if they participate to its activities. The SparkUp startup community organizes events and provides office space for startup teams.

5.2.2.12 University of Vaasa

University of Vaasa has around 5,000 students and 550 faculty members across four schools. It offers two master degree programs: international master's degree program in Strategic Business Development and master's degree program in Management of Growth Businesses. The latter is targeted particularly at individuals who are working simultaneously. Its innovation and entrepreneurship laboratory InnoLab is a phenomenon-based, multidisciplinary open research platform for academics and experts. The Vaasa Entrepreneurship Society organizes entrepreneurship-related events for entrepreneurship-oriented students.

5.2.2.13 Åbo Akademi University

Åbo Akademi University is a Swedish-speaking multidisciplinary university located in Turku, with around 5,500 students and 1,000 faculty members. It offers an entrepreneurship module (20 ECTs) with several entrepreneurship-related courses. This is done through Startup Åbo, which is ÅboAkademi's unit for entrepreneurship. The unit offers support for educators if they wish to develop entrepreneurship-related education. The university has been offering the course, Business Essentials for Scientists, since 2011. It is targeted at faculty members in Åbo Akademi University and the University of Turku which collaborate in promoting entrepreneurship. Åbo Akademi also encourages students to take entrepreneurial courses in other Finnish universities. Its students can participate in the student-led entrepreneurship society Boost Turku, which serves all HEIs in Turku. Similarly, the services of SparkUp are open to students of Åbo Akademi University.

5.2.3 Summary of the Internal Development of EE in Finnish Universities in the 2000s

The policy reports and our investigation of the current state of EE in the Finnish university sector in 2021 provide evidence that EE has developed significantly in the past two decades. In 2005, EE was considered a "relatively marginal phenomenon in Finnish universities," but currently EE is offered at least as individual courses in all universities in Finland. In recent years, the focus of EE has expanded from the strong support of business ownership to intrapreneurship and supporting students' learning. In 2005, universities did not articulate university- or faculty-level objectives of EE, but in 2021 there exists multiple university- or faculty-level strategies for entrepreneurship in line with national and supranational political agendas. Moreover, in many universities, EE has expanded from business schools to the whole university. At the same time, however, some entrepreneurship programs are being discontinued. Many universities have emphasized willingness to develop entrepreneurial activities for students and/or faculty members, and in recent years entrepreneurship studies have become integrated with topical societal phenomena such as sustainability, responsibility, and the arts to address societal changes and challenges. Pedagogical approaches and learning environments have also developed to support the entrepreneurial behavior of students, while education among faculty members has increased. The education of faculty members was a response to the calls of researchers (Béchard & Grégoire, 2005; Ilonen, 2021; Neck & Corbett, 2018) who have put forth teacher education to be an important element of successful EE. Teacher education has been provided as part of formal (e.g., courses, information sessions) and informal activities (informal mentoring, informal networks). Universities are also active in the commercialization of research, and have continued to invest in related support structures and services.

HEIs and other external institutions have increased systematic collaboration locally and nationally in terms of entrepreneurship courses and programs. There are also different support channels for educators who wish to engage in EE. Universities and faculties work increasingly in collaboration with industry, and related models have been developed in recent years. In this regard, universities have caught up with universities of applied sciences.

Entrepreneurship research provides a good basis for EE in Finnish universities. Multidisciplinary entrepreneurship research is conducted in the majority of Finnish universities, and many universities have active research groups with a strong focus on entrepreneurship. Research projects are conducted in collaboration with businesses. In the past decade, student-led entrepreneurship societies have boomed in Finnish university cities to support startup activities and entrepreneurial behavior among higher education students through extracurricular activities. Scholars argue that student-led entrepreneurship societies are important creators of positive energy and thinking around entrepreneurship (Parkkari, 2019; Siivonen et al., 2019). Many universities have started to systematically cooperate with these student-led entrepreneurship societies; in some cases, students can receive course credits by participating in entrepreneurship society activities. EE has reached a relatively comprehensive position amongFinnish universities,but this implies increased complexity due to multiple courses, programs, actors, and collaboration. To summarize, recent EE developments in Finnish universities include:

- EE is currently offered in all Finnish universities.
- University- and faculty-level strategies in entrepreneurship are identified in several universities.
- Multiple approaches and themes of EE: e.g., sustainability, arts, responsibility.
- Increased support of educators.

- Increased collaboration among HEIs locally and nationally.
- Increased collaboration with the business sector.
- Systematic collaboration with student-led entrepreneurship societies.
- Increased complexity due to multiple courses, actors, and collaboration.

5.3 Supporting System of EE

5.3.1 Funding Model of Finnish Universities

Finland's education system is publicly funded. The main financial source is the Ministry of Education and Culture, which also coordinates the activities of universities. The Parliament of Finland annually decides on the amount of core funding that the Ministry of Education and Culture allocates to the universities. This budget is created based on the national financial model. In the 2021 universities' core funding, 42% is allocated based on educational input (numbers of degrees, continuous learning, number of employed graduates, student feedback). In terms of educational input, four percent is allocated based on the number of employed graduates with an emphasis on entrepreneurship. This is closely related to EE, which aims to enhance employability and entrepreneurship post-graduation, and thus encourages universities to promoteEE. 34% of core funding is allocated based on research input, which comprises the number of PhD degrees, publications, and competitive research funding. This encourages high-quality research in any research fields, including entrepreneurship and cooperation with the business sector. 24% of funding is allocated based on other education and science policy considerations. These considerations entail the strategic development and national duties of universities. As entrepreneurship is a strategic element in many universities or faculties, and noted in national educational goals, some additional, albeit minor, funding is allocated based on this. Besides the core funding, universities are encouraged to apply for external financing from sources such as the Academy of Finland, Business Finland, the European Union, foundations, and businesses.

Universities have internal support mechanisms to encourage entrepreneurship and EE. As EE is high in Finland's political agenda, there are several support organizations with which universities can collaborate for promoting EE. The following section will discuss these support mechanisms.

5.3.2 Internal Support in Universities

The core funding model of the Ministry of Education and Culture takes entrepreneurship into consideration in funding allocation, which provides possibilities and incentives for universities to support EE. Internal support can take place at several levels and in different forms either formally or informally. Universities may centrally provide top-down support for EE to encourage staff and students to consider entrepreneurship.

Top-down encouragement of EE implies that the university rectorate would clearly demonstrate its commitment to entrepreneurship on the strategic level and integrate it into everyday activities. This may be done via communication and incentive systems, for example. Internal support may relate to entrepreneurship strategies in universities and faculties, such as in the University of Turku (UTU) (see Sect. 5.4 on UTU Entrepreneurial University). Strategies provide legitimacy for EE within the university. Universities may design internal support services for entrepreneurship, such as staff training or innovation and technology transfer offices. These would generate new research-based ventures as well as coordinate and support funding for entrepreneurial endeavors. University innovation services often manage the existing intellectual property rights portfolio and negotiate sales and licensing agreements. Assessment mechanisms and measures to evaluate the outcomes of EE may be provided top-down at the university level. EE outcomes usually relate not only to education but also to the third mission of the universities, i.e., societal interaction. But these internal support mechanisms are not enough: individuals need to participate in the process (see OECD and European Commission, 2012).

Faculty members take concrete steps to develop EE in a bottom-up approach, whereas encouragement and guidance thereof are provided top-down. It is up to each individual to participate in teachers' training to improve the quality of EE. Furthermore, teachers plan and execute EE courses and programs within the framework of university policy and resources. They may also jointly discuss approaches and, by sharing their experience, learn from their peers. It is important that the university provides forums and opportunities for such activities and articulates their importance, but the very actions are taken by committed teachers. In addition, the university's support structures for commercializing research, for example, are to be exploited by faculty members, and the ways in which this is done depends on the discipline. Technology transfer offices, incubators, and innovation services are also important for graduating students who wish to start new ventures. Such external support organizations are useful, and the following section will briefly discuss these.

5.3.3 External Support Organizations

There are a variety of national, regional, and local organizations dedicated to supporting entrepreneurship in different ways. Locally in different cities there are, for example, informal teacher networks for university and/or higher education teachers where educators can share their experiences and learn from others. These can be particularly important in EE, where experiential learning approach is encouraged. The following lays out some national organizations that the universities can tap on for their EE activities.

Junior Achievement Finland

Junior Achievement Finland positions itself as a leading organization in EE in Finland. It aims to help students learn entrepreneurial and working life skills as well as financial literacy at all school levels, from early childhood to higher education. The organization cooperates with education providers, decision makers, business communities, and students. Junior Achievement Finland offers different study programs that can be integrated into the curricula. The organization equips educators with tools that are crafted in accordance with the current national curricula. One of its best-known programs is a company program where students set up and run businesses for an academic year or semester. The program is widely adopted among HEIs in Finland.

YES Network

YES Network describes itself as Finland's most significant network for the development of EE. It promotes the implementation of EE by strengthening entrepreneurial culture, coaching teachers, developing training, and building school–business collaboration. YES Network organizes several events that support EE, and collaborates widely with different operators and individuals.

The Federation of Finnish Enterprises

The Federation of Finnish Enterprises, SuomenYrittäjät, is a national interest and service organization for SMEs and business owners. Its mission is to improve the position of entrepreneurs and the conditions for entrepreneurship nationally. In relation to this mission, the organization has a strong emphasis on educating educators. The "Competent Teachers—Entrepreneurial Young People" project funded by the Sakari Alhopuro Foundation has contributed significantly to the promotion of EE in Finland. It provides materials, education, and networks for educators who wish or are expected to be involved in EE.

TAT

TAT, Economy and youth-organisation, is a national influencer that aims to help young individuals develop their economic and working life skills and be interested in entrepreneurship. TAT's operations are financed by several Finnish organizations such as the Confederation of Finnish Industries and the Ministry of Education and Culture. TAT provides materials for educators as well as opportunities for students to participate in their programs and events.

Support Materials

Some useful examples of support materials include:

An Entrepreneurship Gamebook,[2] which consists of the joint recommendations of the Rectors' Conference of Finnish Universities of Applied Sciences, Universities Finland UNIFI, and The Federation of Finnish Enterprises for the promotion of entrepreneurship in higher education.

Entrepreneurship for Education Guidelines,[3] provided by the Ministry of Education and Culture, provides information on the principles and concrete measures to promote entrepreneurship and EE at different education levels.

5.4 Characteristics of EE

EE in Finnish universities is governed and guided by the government and its entrepreneurship and educational policies were discussed above. Naturally, these reflect EU-level policies and guidelines, which are further accommodated for regional and local needs. The uniqueness of the Finnish entrepreneurship policy is its cross-sectoral and cross-ministerial approach both at the strategic and operational levels. This implies that ministries attempt to coordinate their individual promotion of entrepreneurship, which indicates a consensual understanding of the holistic nature of the phenomenon of entrepreneurship (Heinonen & Hytti, 2016). The cross-sectoral approach also highlights the importance of entrepreneurship as an engine of economic and social development in Finland.

The leading ministry in entrepreneurship policy is the Ministry of Economic Affairs and Employment. Therefore, it is understandable that the main focus of entrepreneurship policy is not EE as such, but measures to support the development of new businesses and economic growth. However, in the early 2000s, entrepreneurship and EE were high in the entrepreneurship policy agenda to raise awareness of entrepreneurship and to improve entrepreneurial attitudes and skills. Since then, the Ministry of Education and Culture has introduced entrepreneurship to educational institutions at all levels. This is done by integrating entrepreneurship with other subjects such as civic competences in the curricula. Although entrepreneurship is not the focus, many of its facets are implicitly introduced through the curricula. Due to the longstanding work in primary and secondary schools, students' entrepreneurial awareness and attitudes as they enter universities have improved over the years. Therefore, the landscape for university-level EE endeavors has also changed.

Finnish universities have special foci in promoting entrepreneurship through their core tasks in research, education, and social interaction as described in the earlier sections. The focus is on developing knowledge-based academic entrepreneurship and innovation, and multidisciplinarity of universities is considered an asset. EE activities are research-based, though they are implemented in close collaboration with

[2] https://www.unifi.fi/wp-content/uploads/2019/04/Entrepreneurship-recommendations-for-higher-education-institutions-2018.pdf.

[3] Ministry of Education and Culture (2017) accessed at https://minedu.fi/documents/1410845/4363643/Entrepreneurship-for-Education-Guidelines.pdf/ad2a7ecc-ae1b-4460-8fc6-d394e9a0a23a/Entrepreneurship-for-Education-Guidelines.pdf.

stakeholders and embedded in entrepreneurial ecosystems. Student-led entrepreneurship societies play an important role, and these extracurricular activities are typically integrated with entrepreneurship studies.

The governmental push, along with the systematic and long-term measures of the Ministry of Education and Culture, has encouraged universities to introduce and increase their EE offerings and other related activities. The next section highlights an example of an entrepreneurial university in Finland, the University of Turku, which demonstrates the core characteristics of an entrepreneurial university. Many similar characteristics and activities are also found in other Finnish universities.

Example: University of Turku (UTU)—An entrepreneurial university[4]

In 2015, the UTU adopted the "Entrepreneurial University" strategy. This strategy was prepared by a multidisciplinary task force comprising professors, teachers, researchers, and students. In the following year, major elements of the strategy were integrated with the whole University strategy of 2016–2020,serving as a route map for UTU to strengthen its profile as an entrepreneurial university.

The Entrepreneurial University strategy focused on strengthening entrepreneurial attitudes and working practices that support entrepreneurship within the university. Its concrete steps were derived from the Guiding Framework for Entrepreneurial Universities,[5] aimed at European universities and HEIs looking for advice and ideas for the management of institutional and cultural change (OECD & European Commission, 2012). In UTU's strategy, entrepreneurship is promoted mainly through the policy program, "A University for Entrepreneurship,"which incorporates goals for enhancing entrepreneurial attitudes, behavior, and culture across UTU to strengthen the university as an academic and societal influencer. Furthermore, with its multidisciplinary expertise, UTU collaborates withthe business sector and is therefore present in the regional entrepreneurial ecosystem (e.g., Turku Science Park Ltd; startup community SparkUp; and student-led entrepreneurship society Boost Turku). The program comprised various measures with an implementation schedule and involved groups such as the rectorate, deans, staff members, and students. The measures acknowledged the differences of the needs and approaches among faculties and therefore emphasized faculty embeddedness and multi-faculty collaboration. From the beginning, it was decided that the strategy needed to be guided top-down, but implemented bottom-up by the faculties. The entrepreneurial university is coordinated by two designated program managers: one focused on EE, and other on innovation services. They collaborate closely with faculties and external stakeholders. UTU has also nominated entrepreneurship champions and innovation scouts for each faculty.

Next, we look at some exemplary measures that UTU has implemented as an entrepreneurial university. University faculty deans are responsible for implementing the strategy and ensuring that students have opportunities to participate in EE. UTU offers variety of **entrepreneurship studies for students** despite their study program (see also UTU's description in Sect. 5.2 "Internal Development of the EE in HEIs"). All students can take individual courses on entrepreneurship or take entrepreneurship as their minor. The minor degree module, in particular, has become increasingly popular with rising demand from other faculties. Business students can select entrepreneurship as their major at the bachelor's and master's levels and even defend their PhD in entrepreneurship. Some dedicated courses have been designed to support students from different faculties to study entrepreneurship in a multidisciplinary setting:

- *Intoa! Lean Business Program* provides a new type of research and commercialization tool for collaboration with companies. In the program, a multidisciplinary product and business development team—consisting of UTU students and researchers—ideates, conceptualizes, and tests new business concepts for the client company.
- *Startup*! is an 18-week practice-based course organized jointly by three HEIs and a junior achievement program. During the course, students work in multidisciplinary teams and craft business ideas, test suitable business models, and set up businesses that operate in actual markets.
- *Entrepreneurship for Research Professionals Course* is targeted at doctoral candidates at UTU. It fosters multidisciplinarity and the cross-fertilisation of expertise among candidates. The 10-h workshop follows the lean startup method and introduces participants to teambuilding, working with and validating problems, interacting with (potential) customers to get early feedback, developing the business model, and communicating their ideas to different audiences through pitching. The course is run as an intensive, experiential learning workshop.
- *Business Essentials for Scientists* is a joint doctoral course for young researchers at UTU and the Åbo Akademi University. The course includes starting a company, pitching, financing a startup, understanding intellectual property, technology/knowledge transfer, working outside the academic sector, and marketing academic expertise.

The School of Economics also launched the *Entrepreneurship in Residence* (EiR) program, where entrepreneurs participate in the faculty's activities in roles such as guest lecturers, experts, and advisers. EiR is a means to bring entrepreneurs into the classroom and familiarize students with the real life of an entrepreneur and vice versa. Meanwhile, the *Innovation and Business Creation* module provides students with the fundamentals of venture creation and business development, targeted at master's students at the Faculty of Science and

Engineering. In addition to dedicated entrepreneurship courses, UTU provides **entrepreneurship-related training for its educators**, which is an important mechanism in the facilitation of EE. The course *Entrepreneurship and Entrepreneurial Behavior in the University Education*, for example, aims to support the integration and adoption of entrepreneurial pedagogies and an entrepreneurial curriculum, particularly in non-business disciplines.

There are also other means to **encourage the entrepreneurial behavior of staff members, teachers, and researchers** and develop their entrepreneurial knowledge and skills. In order to showcase the entrepreneurial achievements of UTU's staff, the university gives out the annual *Intoa! Entrepreneurial Act of the YearAward* worth €30,000. Two dedicated Entrepreneurship Days are organized around the Intoa! Entrepreneurial Act of the Year Award. Additionally, in 2019, the university launched a small funding for student-initiated entrepreneurship projects in order to support student activities.

UTU has designed an **entrepreneurial path** to enhance the entrepreneurial potential of its students and staff members. The entrepreneurial path comprises education, training, and support services available both in UTU and in the local entrepreneurial ecosystem. The university promotes different services such as startup advisory, co-working spaces, and local governmental business support offices in the region, to guarantee a versatile environment for academic entrepreneurship. The entrepreneurial path and research commercialization processes are further supported by internal guidelines and instructions that help assess, protect, and commercialize innovations, expertise, and skills within UTU.

As a result, the UTU has been able to ensure that students from all faculties can attend at least some entrepreneurship courses. The participation in EE is mainly based on the interest of the students. The majority of the students still come from the School of Economics or other fields where entrepreneurship is incorporated into the degree program. Furthermore, students increasingly participate in student-led extracurricular activities such as Boost, an entrepreneurship society, which are impactful forums for students to acquire practical entrepreneurial skills.

University-level support is important in gaining legitimacy for EE and related activities. Despite its long history in entrepreneurship research and education in the School of Economics and the demand from outside the School of Economics, the incorporation of EE and related activities is challenging. Members from different faculties perceive entrepreneurship differently. We have noticed that, for example, the Faculty of Medicine and Faculty of Science and Technology are more focused on developing business ideas and the commercialization of research findings, whereas the Faculty of Humanities is interested in students' employability. It is understandable that despite university-level strategies in entrepreneurship, the core emphasis of faculty activities is in their respective subject areas and entrepreneurship is considered as complementary. Therefore, the individual's interest defines the level of dedication to entrepreneurship during studies and at work. Furthermore, the strategic interest of UTU in entrepreneurship changes over time, shaping the course offerings and individual interests.

UTU's example of an entrepreneurial university showcases many features that characterize EE in Finnish universities. Students have become increasingly interested in EE: their participation indifferent types of entrepreneurship programs and courses as well as extracurricular activities has grown. This also motivates faculty members and educators to further develop their EE competences and offerings as highlighted by researchers (Béchard & Grégoire, 2005; Ilonen, 2021; Neck & Corbett, 2018). Student-led entrepreneurship societies seem to be active in all cities with university campuses, with inspiring learning experiences and numerous startups created to encourage students and businesses to engage in these activities (Brush, 2014; Pittaway et al., 2010). Indeed, businesses are also becoming more interested in collaborating with universities and different programs to enhance such collaborations, such as UTU's Entrepreneurs in Residence. EE is also closely connected to the entrepreneurial ecosystem, including governmental support agencies, associations, businesses, etc. Hence, EE does not flourish in a vacuum but in close collaboration with other entrepreneurial-minded actors. Furthermore, entrepreneurship studies reflects societal changes: topical themes such as responsibility and sustainability have become integrated into many entrepreneurship courses and programs (Lourenço et al., 2013). Some of the programs and courses may also be targeted at and accommodated for special groups or disciplines such as entrepreneurship for the life-sciences or humanities.

There is a variety of challenges for universities to tackle in order to perform better in terms of EE. The perennial shortage of resources and competences of faculty members is one of them (Liguori & Winkler, 2020). There are some very competent and enthusiastic pioneers in the field of EE, but the numbers are still quite modest. Overall, the skills and competences of faculty members vary significantly. Furthermore, the organization of entrepreneurship programs is spread across universities and faculties without coordination,which makes it difficult to benchmark and learn from others. There are some national and regional competence or

[4] Both of the authors have been involved in developing the University of Turku (UTU) as an "entrepreneurial university." The first author chaired the initial entrepreneurship strategy in 2015, the outcome of which was UTU's Entrepreneurial University Strategy. In addition, descriptions of the practices are based on the confidential self-evaluation report of the university prepared for the Entrepreneurial University Accreditation, which the university received in 2022.

[5] Areas to be covered according to the framework are: (1) leadership and governance, (2) organizational capacity, people, and incentives, (3) entrepreneurship development through teaching and learning, (4) pathways for entrepreneurs (5) university's business/external relationships for knowledge exchange, and (6) the entrepreneurial university as an international institution (OECD and European Commission, 2012).

support centers that provide advice for those interested. Many universities also have dedicated entrepreneurship support services as well as training for staff and potential entrepreneurs. It seems that interested parties have access to such services, but the main challenge is still their limited number, and thus raising entrepreneurship awareness across faculties is required.

EE suffers from some misconceptions and assumptions, which creates challenges. For some, entrepreneurship refers to profit making businesses and therefore they wish to distance themselves from it. Similarly, for some, entrepreneurship evokes negative memories of the great depression in the 1990s when entrepreneurship was introduced as a remedy to a poor employment situation. Later, the notion of opportunity-driven entrepreneurship encouraged universities to support knowledge-based, academic entrepreneurship that relies on one's competences, expertise, and scientific research. On the other hand, some may understand entrepreneurship as the skills and competences needed in working life; they seeentrepreneurship to be synonymous with employability. To put it bluntly, entrepreneurship can be narrowly understood as a moneymaking machine or broadly as anything related to working life skills. This demonstrates the need to critically discuss and understand how EE is understood and what EE needs to deliver in different university settings. Also, the downsides need to be discussed and considered (Bandera et al., 2021; Berglund et al., 2020; Shepherd, 2019). All these assumptions, good or bad, influence the ways in which entrepreneurship is understood and appreciated as well as the ways in which it becomes accepted as a part of university curricula and other activities.

5.5 Toward a Higher-Quality EE

EE in Finland has evolved over time, reflecting economic and societal changes. Its early roots are found in business schools with the focus on the education of economic issues. Later, EE expanded in the 1990s and 2000s when the concept of enterprise education emerged. In the university setting, the development of EE has evolved from raising early awareness to expansion, when an increasing number of universities introduced EE courses for students to be familiarized with entrepreneurship. Currently, all universities offer courses on entrepreneurship, but it seems that a shift in their present and future offerings is already taking place. Based on the information aforementioned, we anticipate at least three alternative futures for EE in Finnish universities. We identify the steps to these alternative futures, and suggest that different universities may diversify in terms of EE and follow different paths.

Running Down Dedicated Entrepreneurship Programs

Some universities run down dedicated entrepreneurship programs and integrate resources with other related disciplines. This may be particularly relevant in business schools, where collaboration with other business studies (e.g., marketing and management, and organization) is evident and a joint playing field may be easily found in the field of innovation, for example. However, this may risk the core of entrepreneurship research gradually deteriorating and fading away, which would pose challenges to EE.

Integrating Entrepreneurship to New Topical Themes and Societal Challenges

Some universities introduce topical themes and phenomenon such as responsibility, sustainability, and digitalization to EE. This is done to bring new flavors to EE and entrepreneurship research. Such phenomenon-based approach may be more impactful in addressing the great challenges in society, but may lose some focus on entrepreneurship.

Strengthening Entrepreneurship Studies with a Multidisciplinary Approach

Some universities continue to invest in the core of EE and integrate the core with other disciplines, so that more students can access entrepreneurship studies. This implies that EE endeavors need to be applied to different disciplinary needs, which requires multidisciplinary collaboration. This poses both challenges and opportunities for actors involved. The approach broadens the understanding of entrepreneurship and provides new avenues for entrepreneurship research and related endeavors.

There may also be other future alternatives that we are not yet able to identify for the Finnish context. We are not suggesting that any of these alternatives are better or worse that the others: they are different. Universities create their own strategies and critically assess how entrepreneurship fits into their overall strategies. We do suggest, however, that it is hard to enhance EE in universities if it is not supported by university-level strategies and top management. Even if these basic requirements—strategy and management—were in place, it takes joint efforts and enthusiasm to implement EE in the university setting. The examples presented in this chapter demonstrate that EE have space and a role to play in the Finnish university sector, although critical assessment is always needed. Indeed, educators need to bring critical and alternative perspectives to EE and not only teach entrepreneurship without pushing students toward criticality and reflexivity (Berglund & Verduijn, 2018). In order to be viable, EE needs to evolve and find new ways of thinking and gaining access to wider audiences. It is a joint and multidisciplinary effort toward a more sustainable and inclusive future for EE.

Acknowledgements The first author would like to acknowledge financial support provided by the NordForskNCoE program 'NordAqua' (#82845).

References

Bandera, C., Santos, S. C., & Liguori, E. W. (2021). The dark side of entrepreneurship education: Adelphi study on dangers and unintended consequences. *Entrepreneurship Education and Pedagog, 4*(4), 609–636.

Béchard, J. P., & Grégoire, D. (2005). Entrepreneurship education research revisited: The case of higher education. *Academy of Management Learning & Education, 4*(1), 22–43.

Berglund, K., Hytti, U., & Verduijn, K. (2020). Unsettling entrepreneurship education. *Entrepreneurship Education and Pedagogy, 3*(3), 208–213.

Berglund, K., & Verduijn, K. (Eds.). (2018). *Revitalizing entrepreneurship education: Adopting a critical approach in the classroom.* Routledge.

Brush, C. G. (2014). Exploring the concept of an entrepreneurship education ecosystem. In S. Hoskinson & D. F. Kuratko (Eds.), *Innovative pathways for university entrepreneurship in the 21st century* (pp. 25–39). Emerald Group Publishing Limited.

Heinonen, J., & Hytti, U. (2016). Entrepreneurship mission and content in Finnish policy programmes. *Journal of Small Business and Enterprise Development, 23*(1), 149–162.

Ilonen, S. (2021). Creating an entrepreneurial learning environment for entrepreneurship education in HE: The educator's perspective. *Industry and Higher Education, 35*(4), 518–530.

Liguori, E., & Winkler, C. (2020). From offline to online: Challenges and opportunities for entrepreneurship education following the COVID-19 pandemic. *Entrepreneurship Education and Pedagogy, 3*(4), 346–351.

Lourenço, F., Jones, O., & Jayawarna, D. (2013). Promoting sustainable development: The role of entrepreneurship education. *International Small Business Journal, 31*(8), 841–865.

Ministry of Education. (2004). *Guidelines for entrepreneurship education and operational programme.* https://julkaisut.valtioneuvosto.fi/handle/10024/80387 [In Finnish].

Ministry of Education and Culture. (2005). *Entrepreneurship as a task of universities?* https://julkaisut.valtioneuvosto.fi/handle/10024/80355 [In Finnish].

Ministry of Education and Culture. (2009). *Directions of entrepreneurship education.* https://julkaisut.valtioneuvosto.fi/bitstream/handle/10024/78869/opm07.pdf?sequence=1 [In Finnish].

Ministry of Education and Culture. (2016). *Good practices of entrepreneurship support in institutions of higher education 2016.* https://julkaisut.valtioneuvosto.fi/bitstream/handle/10024/75399/okm14.pdf [In Finnish].

Ministry of Education and Culture. (2017). *Guidelines for entrepreneurship education.* https://minedu.fi/documents/1410845/4363643/yrittajyyslinjaukset-koulutukseen-okm-2017.pdf/dd81b6e7-888e-45e4-8c08-40d0d5a5277e/yrittajyyslinjaukset-koulutukseen-okm-2017.pdf [In Finnish].

Neck, H. M., & Corbett, A. C. (2018). The scholarship of teaching and learning entrepreneurship. *Entrepreneurship Education and Pedagogy, 1*(1), 8–41.

Organization for Economic Cooperation and Development. (2021). *Inclusive entrepreneurship policies, country assessment notes, Finland 2020.* https://www.oecd.org/cfe/leed/Finland-IE-2020.pdf

Organization for Economic Cooperation and Development & European Commission. (2012, December). *A guiding framework for entrepreneurial universities.* https://www.oecd.org/site/cfecpr/EC-OECD%20Entrepreneurial%20Universities%20Framework.pdf

Parkkari, P. (2019). *Doing entrepreneurship promotion: A critical, practice theoretical study of entrepreneurship* (Doctoral dissertation). University of Lapland. http://urn.fi/URN:ISBN:978-952-337-153-8

Pittaway, L., Rodriguez-Falcon, E., Aiyegbayo, O., & King, A. (2010). The role of entrepreneurship clubs and societies in entrepreneurial learning. *International Small Business Journal, 29*(1), 37–57.

Shepherd, D. (2019). Researching the dark side, downside, and destructive side of entrepreneurial action: It's the compassionate thing to do. *Academy of Management Discoveries, 5*(3), 217–220. https://doi.org/10.5465/amd.2018.0194

Siivonen, P. T., Peura, K., Hytti, U., Kasanen, K., & Komulainen, K. (2019). The construction and regulation of collective entrepreneurial identity in student entrepreneurship societies. *International Journal of Entrepreneurial Behavior & Research, 26*(3), 521–538.

Web-Pages (Accessed in August/September 2021)

Universities in Finland:
Aalto University, https://www.aalto.fi/en
University of Helsinki, https://www.helsinki.fi/en
University of Eastern Finland, https://www.uef.fi/en
University of Jyväskylä, https://www.jyu.fi/en
University of Lapland, https://www.ulapland.fi/en
Lappeenranta University of Technology, https://www.lut.fi/web/en/
University of Oulu, https://www.oulu.fi/en
Hanken School of Economics, https://www.hanken.fi/en
University of Arts, https://www.uniarts.fi/en/
Tampere University, https://www.tuni.fi/en/about-us/tampere-university
University of Turku, https://www.utu.fi/en
University of Vaasa, https://www.uwasa.fi/en
Åbo Akademi University, https://www.abo.fi/en/
Student-led entrepreneurship societies mentioned:
AaltoES, https://www.aaltoes.com/
KuopioES, https://www.facebook.com/KuopioES
JoES, https://joensuues.fi/
JES, https://www.jkles.fi/
LaplandES, https://www.laplandes.fi
LUTES, https://lutes.fi/
OuluES, https://oulues.com/
HankenES, https://www.hankenes.org/
TampereES, https://www.facebook.com/TampereES/
Boost-Turku, https://boostturku.com
Other organisations mentioned:
VaasaES, https://www.ves.fi/
Federation of Finnish Enterprises, https://www.yrittajat.fi/en
Junior Achievement Finland, https://nuoriyrittajyys.fi/en/info/
SparkUp, http://www.sparkup.fi/en
TAT, https://www.tat.fi/
Turku Science Park Ltd https://turkubusinessregion.com/en/
YES Network, https://yesverkosto.fi/en

Chapter 6
Entrepreneurship Education in Germany

Christine K. Volkmann and Marc Grünhagen

6.1 Introduction

Among other movements and institutional initiatives in societies worldwide, entrepreneurship will be pivotal in supporting further social, economic, and ecological development for future generations. Entrepreneurship education (EE) will be key for this development. Themes such as technology, social, and in particular sustainable or green entrepreneurship may provide specific leverage to develop future entrepreneurial and sustainability leaders among the younger generation (Wagner et al., 2021). This is especially true against the backdrop of recent civil movements such as Fridays for Future and Students and Scientists for Future. Since there may be an opportunity to advance EE at German and other educational institutions that encapsulate this spirit and engagement of the young, this chapter will focus on ideas to advance *sustainable entrepreneurship in German higher education institutions* (HEIs).

Generally, there is an obligation to tackle the world's grand challenges, including climate change as well as others pertaining to social welfare, education, environmental protection, food supply, and health, as addressed by the United Nations' 2030 Agenda for Sustainable Development (United Nations, 2015, Resolution No. A/RES/70/1). To transition toward a green economy in Europe, the European Union (EU, 2014) issued the Green Action Plan (GAP), which states that green entrepreneurship should be addressed in higher education to prepare the mind sets of future

C. K. Volkmann
UNESCO-Chair for Entrepreneurship and Intercultural Management, Schumpeter School of Business and Economics, University of Wuppertal, Wuppertal, Germany

M. Grünhagen (✉)
Chair for Entrepreneurship, Innovation and Transformation Schumpeter School of Business and Economics, University of Wuppertal, Wuppertal, Germany
e-mail: Gruenhagen@wiwi.uni-wuppertal.de

X. Xu, *Comparative Entrepreneurship Education*,
https://doi.org/10.1007/978-981-99-1835-5_6

green entrepreneurs. At the institutional level within the education system, in particular, universities and other HEIs play a role in "promoting sustainability principles and thus contribute to the paradigm shift toward more sustainable development" (Tiemann et al., 2018, p. 85; referring to the concept of "sustainable university" in Disterheft et al., 2015). HEIs are essential in this regard because "large-scale and comprehensive societal transitions towards sustainability are considered to be fundamentally knowledge-driven" (Wagner et al., 2021, p. 1141), resulting in the central role of universities and other HEIs in sustainable development within their respective regions (Sedlacek, 2013).

In this three-fold mission of the "third-generation university"—teaching, research, and transfer (Wissema, 2009)—entrepreneurship functions as an essential transfer mechanism of knowledge spill overs. Spill overs may come about in different ways, for example, via university spinoffs from students and scientists, or through the education of future entrepreneurial and sustainability leaders as subsequent alumni entrepreneurs. Correspondingly, our understanding of EE ought to be integrative, including both EE in the classroom as well as coaching, mentoring, and entrepreneurship training opportunities for nascent entrepreneurs and their venture projects. An institutional backbone of education and transfer management in HEIs should support these EE measures as suggested in the holistic framework of EE by Volkmann and Audretsch (2017). The institutional support infrastructure for EE at universities and other HEIs is also vital for backing sustainable entrepreneurs in "solving societal and environmental problems through the realization of a successful business [...] and promoting sustainable development through entrepreneurial corporate activities" (Lüdeke-Freund, 2020, p. 667). At the level of individual entrepreneurs, Wagner et al. (2021, p. 1144) assert that "creating, recognizing and taking advantage of sustainable opportunities are complex challenges...and demand specific support systems." This call for tailored support ecosystems for sustainable entrepreneurs has also been echoed by Volkmann et al. (2021) and Bischoff (2021), as sustainable entrepreneurs cannot establish and grow their businesses without the resources and stakeholder support of those around them (Schaltegger & Wagner, 2011; and generally Stam & van de Ven, 2021). However, in consonance with the verdict of Tiemann et al. (2018), there is still a substantial gap in entrepreneurship research between the contextual ecosystem support of university entrepreneurship on the one hand, and the specific needs of sustainable entrepreneurs and their enterprises on the other hand (see also Volkmann et al., 2021). This chapter aims to contribute to the discussion of *how we can craft and improve sustainable entrepreneurial ecosystem support at HEIs, specifically to promote sustainable EE in Germany, and how to tackle challenges along the way.*

We believe that this perspective offers an interesting view on EE in Germany, particularly because it feeds into the actual societal debate in the country about the need for a broader sustainable transition, for example, in light of the current energy crisis. To shed light on this perspective, the remainder of the chapter is organized as follows. The second section offers a brief overview of entrepreneurship in Germany. Within this vignette, we consider the development of EE orchestrated by different stakeholders in Germany (e.g., education and entrepreneurship policymakers, HEI management, entrepreneurship chairs, and other institutes) from a process perspective of policy streams at different levels. In the third part of the chapter, we zoom in on the internal characteristics of entrepreneurship in German HEIs, presenting an overview of structures and resources for EE at universities and taking a closer look at the core design elements of EE. These include the target groups and outreach of EE across university organizations; EE personnel at HEIs and the formats and contents of curricular and extracurricular EE activities. The fourth section explores the specific case of sustainable EE in higher education, particularly, its supporting entrepreneurial ecosystem. The final, fifth section wraps up the chapter by delving into the nature of sustainable entrepreneurship in HEIs and its possible links to other recent developments such as the digitalization of education prompted by the Covid-19 crisis.

6.2 Entrepreneurship and EE in Germany from a Policy Perspective

6.2.1 Entrepreneurship and EE in Germany: A Brief Overview

While Germany has a strong economy overall in the heart of Europe, the country's entrepreneurial activity, measured by the total entrepreneurial activity rate documented in the yearly international Global Entrepreneurship Monitor study, remains low by international comparison (Sternberg et al., 2021). The level of entrepreneurial activity was higher in the previous reporting period of 2019/2020. However, it resumed its lower long-term trajectory with a reading of 4.8 in 2020/2021, which is similar to the 2018 figure (Sternberg et al., 2021) and in fact remained among the lowest of all GEM countries (see also Metzger, 2021). The recent rate of university spinoff formation and entrepreneurial activity in the context of higher education has been reported to be at around one-per-thousand students in German HEIs (Funke & Schröder, 2020). Overall, it can be said that Germany has a fairly weak entrepreneurial culture as compared to other industrialized, high-income countries (Kalden et al., 2017).

One reason for the comparatively lower entrepreneurial activity in Germany's economy could be related to the strong development of the German labor market over the last decade. This resulted in high opportunity costs of founding activities and a substantial absorptive capacity of employed labor market segments (Metzger, 2021). In addition to the perceived opportunity costs of new venture formation, there are other individual-level attitudes, competence perceptions, and perceived regional opportunities for business formation that may play a role in explaining the lower entrepreneurial activity level (Kaminski, 2016; Sternberg et al., 2021). Typically, the notorious heightened fear and stigmatization of failure may also contribute to this verdict (Kuckertz et al., 2020). Kalden et al. (2017) further found that the economic process of entrepreneurial activity is well accepted in Germany, but a positive connotation of the entrepreneur behind this activity seemingly fails to materialize. Typical weaknesses of German culture for entrepreneurship thus appear to center on deeply rooted social and cultural norms as well as a solid overall economy with a labor market absorbing most of the German working-age population. The attitudes and beliefs on entrepreneurship and on one's capabilities to engage in entrepreneurial activities are consequently an important reason for intensified entrepreneurship and EE policymaking in Germany.

Broadly, EE may be understood as "all content, methods and activities that support the development of motivation, skill and experience, which make it possible to be entrepreneurial, to manage and participate in the value-creating process" (Moberg et al., 2015, p. 14; see also Lackeus, 2015, and Högsdal et al., 2020, for this broader idea of EE including also the formation of entrepreneurial mindsets, e.g., to tackle the world's complex current grand challenges). Within the education system, both HEIs and primary and secondary schools play an important role in further developing this wider definition of EE in Germany. The recent GEM 2020/2021 country report (Sternberg et al., 2021) underscores that improvements have been made in recent years, for example, in establishing the economy (*Wirtschaft*) as a subject in school curricula. However, progress has been fairly fragmented due to the federalist structure of German schools and the higher education system with its decentralized state (*Länder*) responsibilities for education policy.

Similarly, the focus of public tertiary education in Germany (and EE within the HEI system and education policy) is characterized by this decentralized structure. HEIs (*Hochschulautonomie*) have substantial responsibility and autonomy in terms of participating in the global competition within the education and innovation sector (Hochschulrektorenkonferenz, 2021). This independent mandate of HEIs embraces education and science management for crafting personnel and other infrastructure for entrepreneurship training and support, as well as the autonomy to design research and teaching, guaranteed under German constitutional law. Generally, the fundamental organizational structures of HEIs in Germany is an element of the state legislation, and HEIs are tasked with the subsequent implementation and operational design (Hochschulrektorenkonferenz, 2017). Within this regulatory framework, there is a range of country-wide federal initiatives to promote entrepreneurship and EE (e.g., the EXIST program, which will be discussed below), as well as EE policies and university entrepreneurship programs at the regional state level (e.g., the Exzellenz Start-up Center initiative in NorthRhine-Westphalia, Germany's most populated state). Principally, EE initiatives in German HEIs—including those aimed at supporting education for sustainable entrepreneurship—unfold against the background of this institutional and legal framework, competing for resources and material infrastructure with other science disciplines in university organizations. It is therefore critical to reflect on how sustainable EE can rank higher on the entrepreneurship and education policy agenda, and how to improve the infrastructure for entrepreneurship teaching and coaching. More extensive support for sustainable EE in HEIs may also benefit the further development of EE in university organizations overall, as sustainable venturing for the benefit of society may be more compatible with academic values and norms than outright for-profit growth entrepreneurship (cf. Wagner et al., 2021). The following section introduces a perspective of sustainable EE as a process of policy streams and agenda-setting.

6.2.2 Developing Education for Sustainable Entrepreneurship as a Process of Multiple Policy Streams

It has been argued that sustainable entrepreneurs may have a central role in leading sustainability transitions (Hörisch, 2015; Wells, 2016). Multilevel sustainable transitions can be described as the transition of socio-technical systems as a complex and profound change across different levels (Geels, 2011). Such systemic transitions arise from the coevolution between economy, society, and ecology (Loorbach & Wijsman, 2013). In transition research, the levels of these transitions are framed as regime (at the top), across landscape, and niche (at the bottom). In particular, sustainable entrepreneurs initiate ventures in a local geographic niche (e.g., within a university region, starting their venture from campus). Characteristically, sustainable entrepreneurs (and other actors such as eco-activists) are bound by institutional and societal structures, especially at the landscape and regime levels, inhibiting swift changes toward greater sustainability. It is precisely this tension that sustainable entrepreneurs address, for example, an ecological or social problem in their initial niche. In addition, widening the scope of a new product or service solutions innovated by sustainable entrepreneurs will require further top-down support from important resource providers and policymakers so as to reach a regime-wide impact (e.g., in a traditional industry). One issue in this movement will be the focus on EE and training for students of sustainable entrepreneurship as well as coaching and funding concrete sustainable venture projects of green entrepreneurs from academia. These could be via university-level education policy or external policy support programs for sustainable entrepreneurship. In essence, this will also be a process of political agenda-setting. While the discussion in this chapter is not fully-fledged comparative policy research, it is still worthwhile exploring the potential for further developing EE for sustainable entrepreneurship in Germany against the background of a rather fragmented education and science policy system with resourceful actors. These actors—such as federal and state-level ministries, university management and rectorates, and local policymakers in university regions—define the resource provision and network support for the work of entrepreneurship chairs, institutes, and centers that deliver entrepreneurship training and coaching at the front end within HEIs.

Kingdon's (1984) framework of multiple policy streams is a useful tool for this reflection, since it provides a process theory for political agenda-setting alongside three different streams in complex, fragmented political systems. These are the *problem, policy, and political streams* (Béland & Howlett, 2016):

- The *problem stream* encapsulates issues considered as public problems that the government should address. Political awareness of these problems arises in different ways, for example, through acute crises such as the Covid-19 pandemic, through feedback loops triggered by social movements like Fridays for Future, or by sustainable entrepreneurs who tackle a specific societal or ecological challenge.
- The *policy stream* consists of experts such as scientists who provide potential solutions and measures to cope with a problem such as global warming or social exclusion. These suggestions in turn provide a pool of possible policy actions that are evaluated and selected from the process of political opinion formation.
- The *politics stream* comprises more general aspects of political institutions that shape the ultimate political agenda-setting and action, such as executive and legislative changes or broad transformations in public opinion.

The role of sustainable EE in all this lies mainly in educating mindsets for sustainable entrepreneurship, and supporting sustainable entrepreneurs from university who start from their local ecosystem niche and contributing to the problem stream. In addition, universities and other HEIs serve as input providers to the pool of possible solutions for ecological and other sustainability challenges through their research within the policy stream. Innovative solutions from research and development also constitute opportunities for sustainable entrepreneurial venturing as one way to transfer and diffuse scientific discoveries into society (e.g., renewable energies or alternative drive systems). In view of the current pressing need to address climate change and to further sustainable development, there may be scope for what Kingdon (1984) framed as *policy windows* for political action in situations where the three streams cross and define policy agenda, be it temporary (cf. Béland & Howlett, 2016). The scope for such a policy window of possible catalysts (and barriers) for more sustainable EE to be on the agenda of education policymakers and university managers has also been addressed in the discussion of multistakeholder governance in entrepreneurship (Wagner et al., 2021). This has been particularly addressed in the form of governance and support structures for sustainable entrepreneurs in sustainable entrepreneurial ecosystems and EE ecosystems (Bischoff et al., 2018; Volkmann et al., 2021; also see the following sections).

Any exploration of support infrastructures for sustainable or other forms of entrepreneurship in regional ecosystems needs to consider the sociocultural and economic context (Pankov et al., 2021; cf. Welter et al., 2019). In Germany's case, support for sustainable entrepreneurs and their education evolves in the context of German education and entrepreneurship policy and within the institutional structures of the decentralized higher education system. The following sections will elaborate on and discuss selected aspects of education for (sustainable) entrepreneurship in Germany, policy programs that promote entrepreneurship, and the role of institutional factors of HEIs within the framework of sustainable entrepreneurial ecosystems.

6.3 EE at German HEIs: Resources and Core Elements of EE

6.3.1 Structure and Resource Base for EE in German Higher Education

Principally, public higher education in Germany is structured into two main types: universities and universities of applied sciences (*Fachhochschulen*). The former is more research-oriented but also has teaching and transfer on the agenda, which is relevant for a hands-on, practical discipline like entrepreneurship. As the name suggests, the latter is more focused on applied forms of academic work. However, both types of HEIs have increasingly engaged in EE over the last two decades, with a rising number of entrepreneurship chairs and institutes (Funke & Schröder, 2020). Most of these organizations are publicly funded with a comparatively much smaller number of private institutions catering for specific academic disciplines and target groups.

In terms of resources and financing, publicly funded HEIs also rely substantially on external third-party funding (*Drittmittel*) in addition to their public budget. In recent years, the trend toward tertiary education has led to a substantially growing enrollment above and beyond the rise in the public budget for HEIs (OECD, 2020). Considering this general tension of university funding lagging behind organizational growth, entrepreneurship at German HEIs has and will have to compete for financing and other resources both with regard to public baseline funding (e.g., whether faculties and rectorates decide to sponsor an additional entrepreneurship faculty) and external third-party funding (e.g., competing with other science disciplines for endowments). A recent large-scale study with more than 200 participating German HEIs on their infrastructure of entrepreneurship reported that institutions had around 0.5% of their overall budget (baseline budget and third-party funds) available for entrepreneurship (Funke & Schröder, 2020). Interestingly, in 2019, about two-thirds of this total amount of financing came from temporary, often project-based, third-party funding, i.e., around 90 million euros of a total of approximately 130 million euros (Funke & Schröder, 2020). This predominance of temporary external funding for EE operations at HEIs seems unsurprising given the substantial number and funding volumes of public and private initiatives in entrepreneurship and EE policy promotion in Germany. For example, the Exzellenz Start-up Center program of the North Rhine-Westphalia State Ministry of Economic Affairs, Industry, Climate Action and Energy alone provides 116 million euros to six universities in the 2019–2024 funding cycle.[1]

However, the most influential entrepreneurship policy in German higher education has been the federal-level EXIST initiative, which has shaped the landscape of EE in German higher education and research for more than two decades. The EXIST initiative demonstrates some crucial characteristics of EE in Germany, including both

[1] https://www.exzellenz-start-up-center.nrw/.

catalysts and barriers. The *EXIST—Start-ups from Science Programme*[2] originated in 1998 and is currently in its fifth phase (EXIST V). "Initiated and run by the German Federal Ministries of Education and Research and, later, Economics and Technology, it covers a wide range of support areas, in particular awareness raising, entrepreneurship education, generation of business ideas from science and research, consulting and counselling, financial and infrastructure resources and the overall development of academic entrepreneurial culture" (Volkmann & Grünhagen, 2014, p. 232; cf. Kulicke, 2014). The different phases of EXIST concentrate on various aspects of the above-mentioned areas of support for entrepreneurship training and spinoff formation, mostly from German universities but also research institutions such as the MaxPlanck Institutes (Volkmann & Grünhagen, 2014):

- EXIST I "model regions" (1998–2001; funding of 14 million euros): established an initial infrastructure for university entrepreneurship through a network approach in five university regions in Germany.
- EXIST II "transfer" (2002–2004/2005; funding of 11 million euros): supported knowledge transfer to other university regions in Germany to further develop entrepreneurial infrastructure at more HEIs.
- EXIST III "specific projects" (2006–2011; funding of 40 million euros): follow-on funding of existing infrastructure frameworks for entrepreneurship and of specific support instruments such as university incubators or startup centers.
- EXIST IV "entrepreneurial universities" (2010–2018; funding of 46 million euros): support for the establishment of a culture of entrepreneurship across HEIs, in particular, outside business and economics departments.
- EXIST V "potentials" (2019/2020–2024; funding of approximately 150 million euros): nation-wide leveraging of opportunities for science entrepreneurship in around 100 HEIs with a focus on regional networks and internationalization.

In addition to providing funding for HEIs and research institutions to build and complement the entrepreneurial infrastructure in their institutions and provide EE to students, EXIST offers funding and coaching to individual teams of nascent academic entrepreneurs during the seed phase of their new venture projects—the Gründer-Stipendium and Forschungstransfer grants. The latter particularly funds high-tech startup projects by researchers and scientists at HEIs and research institutes.

In summary, the EXIST program has helped substantially in initiating and establishing a resource and personnel base in terms of both EE (e.g., by co-funding initial entrepreneurship chairs at German public universities) and academic spinoff support. This politically initiated top-down path differs from the evolution of entrepreneurship at Anglo-American universities such as Cambridge or Stanford. There, entrepreneurship has evolved and grown over many years from within university campuses via the engagement of individual academics, alumni startups, and expanding entrepreneurial networks in the corresponding university regions. In contrast, in Germany, academic entrepreneurial activities and EE were fairly limited at the turn of the millennium when EXIST commenced (Grünhagen et al., 2005). The early phases of

[2] www.exist.de

EXIST allowed German HEIs to experiment with and explore different formats and approaches toward EE and scientific startup support. However, it also became clear that there would be no one-size-fits-all concept for entrepreneurship support, bearing in mind the heterogeneity of autonomous German HEIs with different disciplinary foci and institutional size (Volkmann & Grünhagen, 2014).

Finally, the characteristics of the EXIST program, specifically such a form of political initiation of entrepreneurship at HEIs, can be seen in the rather artificial, politically staged nature of developing entrepreneurial spirit during the program's different phases (ibid.). In particular, temporary, project-based funding via third-party endowments, including EXIST, has been considered a problem (cf. Kulicke, 2018). On the one hand, such temporary subsidies seem compatible with incentive structures that encourage universities to provide internal funds to finance entrepreneurship activities in the long term (e.g., EE lecturing and startup consulting personnel, infrastructure for entrepreneurship centres, etc.). On the other hand, temporary funding practices risk the end of entrepreneurship support and training activities when such funding ends. This situation may have led universities to request the state to devote more resources to EE and entrepreneurship support for the longterm as part of its permanent core funding, so as to ensure the continual running of entrepreneurship activities (Funke & Schröder, 2020). This call for continuity in entrepreneurship and EE from German HEIs highlights a possible tension between the predominant temporary, short-to-medium-term, external funding of EE and the desire to have entrepreneurship offerings fully institutionalized and internally funded from regular university budgets on a permanent long-term basis. This *potential imbalance of funding long-term EE operations with rather short-term, often project-based, external funding* is also essential in view of the complex and differentiated core structures of EE at German HEIs, which will be described briefly in the next section.

6.3.2 Core Elements of EE at German University Organizations

From a bird's-eye perspective, EE at HEIs is constituted by three interrelated pillars: *institutional aspects of EE*, *curricular EE offerings*, and *extracurricular EE activities.* We will address selected aspects of these pillars via a holistic framework of EE (Volkmann & Audretsch, 2017).

- *Institutional aspects of EE*: university's entrepreneurship and EE strategy and commitment; development of EE-related organizational units and staff positions; integration of EE across the university; regulation and incentive structure for EE; development of entrepreneurial mindsets of university members.
- *Curricular EE offerings*: core design of EE courses, modules, and study programs (format, content, teaching methods, evaluation, and feedback); interdisciplinary integration in degree programmes across all faculties.

- *Extracurricular EE activities*: range of interactive, practical EE formats; cooperation with external stakeholders from the entrepreneurial ecosystem; core design of extracurricular EE; coordination with curricular offerings; and planning of curricular integration.

Within the core development of EE, the following aspects of EE management are relevant for German HEIs across curricular and extracurricular EE (Volkmann & Audretsch, 2017):

- target groups of EE (including EE offerings outside the economics faculty)
- personnel for EE units (personnel resources, institutional affiliation, professional development)
- core EE content offerings (content, format, and methods)
- support for students (coaching and further support for entrepreneurship students and academic entrepreneurs)
- quality assurance (development of EE and curricular integration in accredited study programs).

For *target groups* of EE at German HEIs, the most important issue is improving the outreach of EE beyond universities' business and economics faculties and departments. EE is often integrated within economics, and most entrepreneurship chairs were traditionally established in business and economics faculties (Kulicke, 2018). However, university-wide, interdisciplinary expansion of EE has developed over the past decade, both internally and as a deliverable in third-party sponsored projects relating to university entrepreneurship. Funke and Schröder (2020) reported that the majority of curricular EE offerings are still targeted at students of social sciences and economics, while many EE offerings are domiciled in engineering faculties. There were fewer EE courses in the humanities and liberal arts faculties. One reason for this may be that the structures of organizing university-wide EE at German HEIs predominantly follow a "magnet model" of centralized entrepreneurship centers. This model focuses on extracurricular EE offerings rather than curricular courses, as opposed to a "radiant structure," which would have curricular EE courses in every faculty (cf. Funke & Schröder, 2020; Volkmann, 2009).

Most *entrepreneurship and EE personnel* at German HEIs are domiciled at entrepreneurship chairs/institutes and entrepreneurship centres. Currently, there are around 190 entrepreneurship and entrepreneurship-related chairs and professorships: 60percentinuniversities and about 40% in universities of applied sciences. According to a German research institution for entrepreneurship, the overall number grew over the last two decades from around 20 in the early 2000s to 100 chairs in 2010 and more than 150 in 2020. Typically, these entrepreneurship chairs are home to a number of affiliated doctoral and postdoctoral research employees. These employees would be active in EE teaching and entrepreneurship research on a temporary contract basis, which is common among German HEIs. Additional personnel in other entrepreneurship units such as entrepreneurship centres, university transfer offices, or centralized advisory offices for startup consulting, are often also employed on a temporary basis due to the aforementioned fixed-term third-party projects. The prevalence of a substantial number of temporary and often part-time personnel in EE with the corresponding fluctuation will likely impact the professional development of EE personnel, which is instrumental to advancing entrepreneurship teaching methods and formats (EU, 2015; Volkmann & Audretsch, 2017). The need to develop the didactic know-how of lecturers became even more important during the Covid-19 pandemic, when most entrepreneurship courses had to be shifted online, which required lecturers to develop skills in online EE teaching (Liguori & Winkler, 2020; Ribeiro et al., 2020).

Correspondingly, the pandemic changed how *core EE offerings* are delivered at German HEIs. However, this also provides future opportunities for thorough further digitalization and expanded online outreach as higher education moves from emergency remote teaching to proper digital education (EU, 2021). Regarding the contents of EE in Germany, Funke and Schröder (2020) reported that EE courses predominantly focus on the managerial basics of entrepreneurship and startup management, developing entrepreneurial opportunities or business models, and design sprints. Less common are courses covering prototyping or intrapreneurship. Apart from these, an interesting feature is the growing segmentation of EE course contents as compared to the early days of EE in German HEIs. Typical for this segmentation are EE courses that focus on areas such as social, sustainable, and female entrepreneurship (Klusmeyer et al., 2015). This segmentation may, however, be beneficial for the further differentiation of course offerings to students from different disciplines, who may find novel forms of entrepreneurship aimed at solving societal problems more attractive than for-profit business entrepreneurship. This will also be important to improve education for sustainable entrepreneurship, as the case study in the following sections shows. As far as formats are concerned, beyond the Covid-19 pandemic shift, typically seminars and workshops dominate, followed by lectures, project work, and pitch events. Less common are more specific and elaborate formats such as entrepreneurial business simulations, hackathons, or summer schools (Funke & Schröder, 2020).

Traditionally in German HEIs, EE that is directed at larger audiences (e.g., lectures and events) has been accompanied and augmented by individual coaching and consultation *to support university members* (students, researchers, alumni) who are nascent entrepreneurs and plan to start their own venture. In the past, entrepreneurship chairs offered startup consulting, but it is now more common to see centralized entrepreneurship centers and campus-wide transfer units take on this role. The level of resources for this individual support of academic entrepreneurs and the overall infrastructure of centralized units such as entrepreneurship centers depends substantially on the support and benevolence of university top management. For most HEIs in Germany (reportedly more than 80%; Funke & Schröder, 2020), fostering entrepreneurship is a public obligation that is often included in so-called "target agreements"(*Zielvereinbarungen*) with state ministries, or a part of the university's individual transfer strategy. Less common are explicit, institutionalized strategies for promoting startups (Funke & Schröder, 2020).

The future *quality assurance of EE* will also evolve based on the universities' entrepreneurship strategies and approaches concerning: (1) a further digitalization of education, (2) the employment of temporary versus permanent EE personnel, and

(3) the future role of transfer activities outside the curriculum and corresponding curricular integration. On the one hand, such extracurricular activities entail ample opportunities for networking and cooperation with external stakeholders such as regional entrepreneurs, incubators, regional technology centres, startup consultants, and other professionals in entrepreneurial management. Also, extracurricular activities are often easier to organise compared with placing novel EE curricular offerings in accredited degree programs. Therefore, they serve as a suitable pathway for getting entrepreneurship started at HEIs that are new to the subject (EU, 2015). On the other hand, however, extracurricular activities are far less institutionalized, running the risk of being discontinued in situations of budget constraints, personnel changes, or crises such as the pandemic. Additionally, though they often providing excellent learning experiences, extracurricular activities suffer from a lack of credit incentives (EU, 2015).

Notably, the majority of EE offerings at German HEIs today are extracurricular. The total number of EE courses and programs grew from merely 100 in the early 2000s, to approximately 250 in 2007 and finally to more than 7000 in 2019. Of the last figure, 3600 were curricular and more than 3800 were extracurricular (Funke & Schröder, 2020). This development of extracurricular offerings becoming even more common than institutionalized curricular offerings may have been reinforced by the many third-party funded university entrepreneurship projects. These often demand HEIs to engage in practical, hands-on, and interactive EE teaching in cooperation with external stakeholders to increase the output of spinoffs. For university management and EE leaders in HEIs, this focus on extracurricular EE results in developing the quality and perpetuating successful extracurricular activities. Hence, one option to improve the institutional character of selected EE offerings in the long-term is to consider their curricular integration (EU, 2015).

The above-mentioned strategic aspects and the crafting of EE in general at German HEIs constitute the environment in which relatively new formats of EE such as sustainable entrepreneurship operate. To this end, the following will be key: approaches to attract student target groups outside the business faculty to be interested in sustainable EE; modes of collaboration with various external stakeholders in university regions which are engaged in sustainable development and transition and solid concepts of practical extracurricular activities in sustainable entrepreneurship (which seems particularly suitable for voluntary project-based engagement).

As regards the future strategic evolution of German higher education in terms of the use of technologies and the resource base of EE in Germany, the role of the fourth industrial revolution as a more general shift within the socio-technological context of education in HEIs will be important. Of course, concrete impacts of this revolution, also known as industry 4.0, are yet to be seen, e.g., concerning the integration of artificial intelligence (AI) in teaching in higher education. However, at the same time, already today the covid-19 pandemic has prompted a wave of novel digital technology use, both in EE and within the creation and adaptation of business models of startups. It is thus worth to briefly touch upon two intervened fundamental aspects of the possible impacts of the fourth industrial revolution on EE: first, potential changes in "what we teach", in particular around future digital, industry 4.0-based, business models and novel digital entrepreneurial competences around this, and second, the possible ways in which the communication and delivery of EE itself as to "how we teach" may develop around novel digital technologies.

"Digital technologies are transforming the nature and scope of entrepreneurial activity" (Chalmers et al., 2021, 1028). Foremost, this will require further developing the teaching of digital business model development in German EE. Since the technologies used in this domain are complex and often very advanced, this may not be taken on board across German HEIs and their faculties with the same speed. Likely, technical universities and universities of applied sciences with strong engineering and science departments may take the lead towards this end (in correspondence to more general technology entrepreneurship which also appears to be geared more towards larger, technical university institutions; Funke & Schröder, 2020). In view of issues such as interconnectivity within, e.g., the internet of things (IoT) or human–machine interaction, which may well be relevant overall to business and industry overall in the future, there may also be growing integration of entrepreneurship and (small) business management issues with the field of innovation management. For entrepreneurship, this may both involve creating novel digital business models as well as serving for digital business model transformation for larger enterprises (e.g. within the strong German SME "Mittelstand" sector) where knowledge-intensive start-ups may offer novel smart service and other technology-based business-to-business solutions. At the level of individual entrepreneurship students, German EE will have to build novel competences for digital business model creation and digital start-up management practices using digital tools (such as in agile project management). In particular, German HEIs will require to improve the innovation capacity and output of innovation and entrepreneurship students in a structured way such as sketched out in the competence framework for entrepreneurs within the fourth industrial revolution by Blignaut and Botha (2022). Such competence-based EE concepts may be linked to other digital competence frameworks, e.g., the more comprehensive DigComp framework of the EU. The verdict advocated by the OECD (2018) that this will mean to prepare the young generation of students for the unknown in many ways (e.g., in terms of future technology use and job profiles) is pivotal here. This is because "preparing for the unknown" will imply that German HEIs will require to be flexible in their approach as to what contents of EE and the technologies around them they integrate in their EE curricula. Likely, such a development will come about in many different small steps in which new experimental EE formats, which address single (or a small number) of competences, evolve at individual university organizations and will diffuse into the wider university landscape subsequently. Correspondingly, such an emergent development of new EE formats addressing different aspects of the fourth industrial revolution will require substantial flexible further training of EE instructors and lecturers in the future. The recent covid-19 pandemic has prepared the ground for intensified technology use in university teaching of EE (and generally), including "train the trainer" initiatives to prepare university staff to accommodate to the world of digital and remote EE teaching. One key aspects in this regard will be the linking of technology use around the principles of the fourth industrial revolution in both EE teaching and the nature of digital business model development. This will be important not only for (pure) technology entrepreneurship in itself, but also for technology-based solutions in other forms of entrepreneurship such as social or sustainable entrepreneurship where technology may also provide sustainable solutions for societies' ecological and other problems.

6.4 Supporting Ecosystem Elements for Sustainable EE

6.4.1 Sustainable Entrepreneurship and Entrepreneurial Ecosystems

In the introduction, we have characterized *sustainable entrepreneurship* as addressing social and ecological issues via business formation (c.f. Lüdeke-Freund, 2020). The strengthening of sustainable entrepreneurship in HEIs is important because the resolution of environmental and social problems by doing business the green way (Gast et al., 2017) also encapsulates "development gains for others"(Patzelt & Shepherd, 2011, p. 632).

This specific nature of sustainable entrepreneurship in terms of the benefits and positive externalities for others in society will require further attention in sustainable EE, since the distinct characteristics of sustainable entrepreneurship invite additional challenges for sustainable entrepreneurs to establish and grow their ventures. DiVito and Ingen-Housz (2021; also see Fig. 6.1) put forth three central challenges for sustainable entrepreneurship:

- *Distribution of benefits*: Compared to traditional for-profit business entrepreneurship, sustainable venturing will produce more collective benefits for society. This makes the internalization of values more difficult for sustainable entrepreneurs (York & Venkataraman, 2010), including the acquisition of an external resource base, in particular, funding.
- *Prolonged duration for the establishment of sustainable ventures*: Since sustainable entrepreneurs aim at solving grand challenges (e.g., providing alternative renewable energy sources), they may need more time to build and grow their organizations. Correspondingly, this may necessitate the establishment of more long-term relationships with patient and trusting external stakeholders before the ventures bearfruit. The role of such ecosystem stakeholders may be instrumental, especially at the start, as they provide initial legitimacy (Kuratko et al., 2017) and allow sustainable entrepreneurs to build a tangible showcase of their vision to convince more reluctant stakeholder groups subsequently.
- *Unique composition of stakeholders in the sustainable entrepreneurial process*: The process of recognizing and, in particular, exploiting sustainable entrepreneurial opportunities may require a different set of actors as compared to traditional entrepreneurship (DiVito & Ingen-Housz, 2021). For example, at the outset, eco-activist groups may be essential to raise a novel ecological issue to

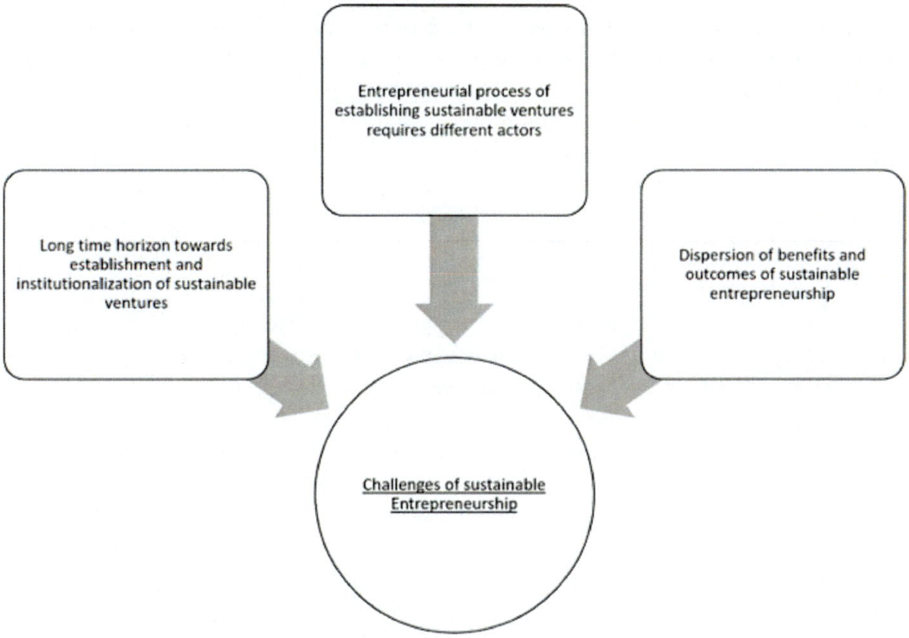

Fig. 6.1 Challenges of sustainable entrepreneurship. Adapted from DiVito and Ingen-Housz (2021)

be acted upon. Subsequently, addressing this issue may need innovation efforts by entrepreneurs and technology partners. "[T]he recognition and exploitation of sustainability opportunities may require different actors—those that highlight the issues, those that invent alternative products or materials and those that take entrepreneurial action" (DiVito & Ingen-Housz, 2021, p. 1064).

As sustainable entrepreneurs must meet the above challenges and particularly navigate the trade-offs between external social and internal economic sustainability (Hahn et al., 2015), their entrepreneurial journey may be more complex as they compete against for-profit rival businesses (Patzelt & Shepherd, 2011). Correspondingly, sustainable EE must assist future sustainable entrepreneurs and other such leaders in tackling these challenges. Also, the role of universities will involve creating more future sustainability leaders who provide different possible solutions to socio-ecological problems, since we do not know ex ante what will and will not work, for example, in combating climate change. Tiemann et al. (2018, p. 85) therefore propose that universities sensitize and educate "future sustainable entrepreneurs and use a systematic approach to specific opportunities and challenges of sustainable development by providing appropriate support systems." These support systems provide the context in which sustainable entrepreneurship at HEIs may flourish, underscoring the role of sustainable entrepreneurial ecosystems in university regions.

Support for sustainable entrepreneurs and their ventures at the local or regional levels is critical. This is because sustainable ventures cannot be built in a vacuum without resources from stakeholders who provide, for example, finance, knowledge, technologies, human capital, and legitimacy (Schaltegger & Wagner, 2011; Stam & Van de Ven, 2021). At the same time, it is not perfectly clear yet in entrepreneurship research what exactly sustainable ventures need compared to other startups (Volkmann et al., 2021). However, there is the idea that sustainable entrepreneurs will require tailored support (Bischoff, 2021) and that, principally, "sustainable entrepreneurs may require different ecosystems where actors interact and provide support in significantly different ways than in traditional entrepreneurial ecosystems" (DiVito & Ingen-Housz, 2021, p. 1058).

The need for flexible support for sustainable entrepreneurs also shows the importance of *supportive contextual policy streams in the form of prospering sustainable entrepreneurial ecosystems* at the local/regional niche level, the surrounding landscape, and the national regime-level policy support. And since the resource and know-how needs of sustainable ventures change, being unclear at origin, and the architecture of environmental support evolves over time, entrepreneurial ecosystems are best understood as "a [*dynamic*] community of interdependent actors [...] and system-level institutional, informational, and socio-economic contexts" (Audretsch & Belitski, 2016, p. 4). Sustainable entrepreneurial ecosystems, in particular, may be conceived as "an interconnected group of actors in a local geographical community committed to sustainable development through the support and facilitation of new sustainable ventures" (O'Shea et al., 2021, p. 1097).

Concerning the support needs for the facilitation of sustainable ventures, DiVito and Ingen-Housz (2021) connect the challenges faced by sustainable entrepreneurs introduced above (a long time horizon for establishment; different composition of support actors; implications of dispersed benefits of sustainable entrepreneurship) with *conditional aspects of sustainable entrepreneurial ecosystems* (see Fig. 6.2).

- *Actors' orientation toward sustainability*: A core condition for a functioning ecosystem support may be a shared interest or even a vision related to sustainability, and an emergent supportive emotional climate with actor communities as "meta-enablers" for "a sustained engagement with the ecosystem beyond an individual's efforts at venture creation" (O'Shea et al., 2021, p. 1099).
- *Sustainable opportunity recognition and subsequent resource mobilization*: Within sustainable entrepreneurial ecosystems, actors and processes that highlight ecological and social problems (e.g., by activist groups) will be key and should feed into subsequent active resource mobilization aimed at solving these problems, inter alia, by sustainable venturing. This resource support will be critical for increasing the impact of such initial niche solutions to grow to a larger scale.
- *Collaborative, sustainable innovation*: The willingness and scope for collaboration in developing novel products and service solutions sustainably to replace non-sustainable offerings will be instrumental, since such products and services may be the nucleus of business models of sustainable ventures.
- *Regional demand for sustainable goods and services*: At the other end, an ecosystem can and should also function as an initial demand source for novel

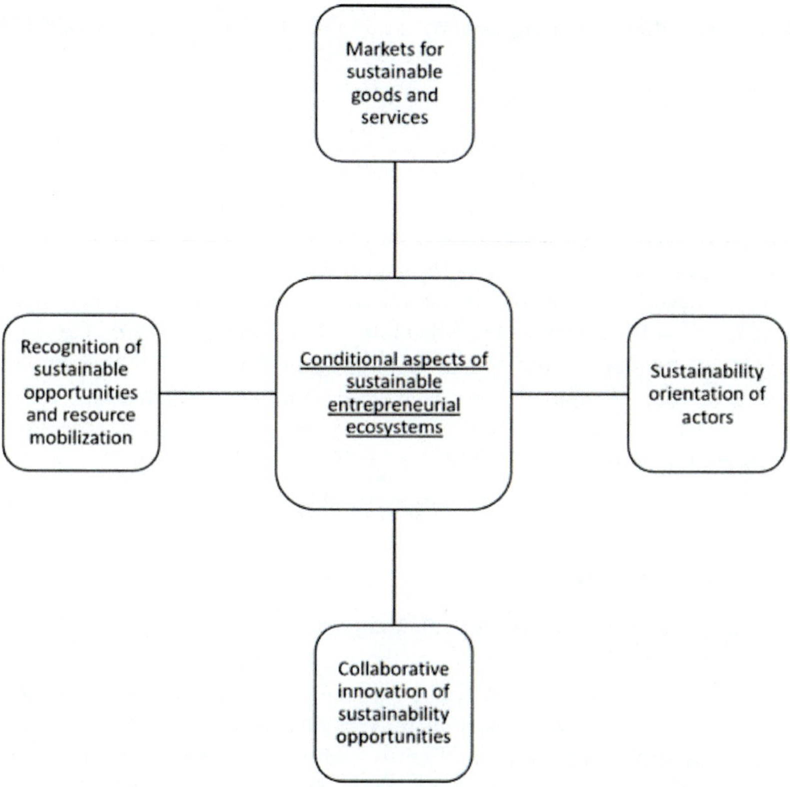

Fig. 6.2 Conditional aspects of sustainable entrepreneurial ecosystems. Adapted from DiVito and Ingen-Housz (2021)

products and services, allowing sustainable entrepreneurs to experiment with early-stage versions of their product offerings and build successful showcases to grow beyond the local ecosystem niche.

Such sustainable entrepreneurial ecosystems see actors share a sustainability orientation, identify ecological and social problems together, collaborate on developing sustainable product offerings, and allow sustainable venture projects to grow from a regional demand base. These constitute a niche environment where increases in entrepreneurial skills and capabilities of versatile support actors can produce and scale economic and social benefits (Greene et al., 2010). A key factor is university-related entrepreneurial ecosystems, which generates such increases in entrepreneurial skills and capabilities in higher education. Educating for entrepreneurship in general, and sustainable entrepreneurship in particular, may be considered an integral part of such university-based entrepreneurial ecosystems (cf. Regele & Neck, 2012). The next section takes a closer look at some elements of ecosystems for sustainable EE in the German HEI context.

6.4.2 Elements of Supportive Ecosystems for Sustainable EE

The elements of *internal* entrepreneurial ecosystems in HEIs comprise curricular and extracurricular EE offerings as well as entrepreneurship research and evaluation measures (e.g., for the impact of entrepreneurship programs or the success of venturing processes) (Greene et al., 2010). These elements for delivering high-quality EE programmes in sustainable entrepreneurship relate more to the external ecosystem elements of stakeholders who provide various forms of support alongside challenges faced by sustainable entrepreneurs (cf. Fig. 6.1). Additional typical and principal *external* ecosystem elements include the social network of actors providing material (e.g., financing) and immaterial (e.g., know-how) resources as well as human capital (e.g., future employees in sustainable ventures) (Kansheba & Wald, 2020). The efforts of HEIs to build and maintain a university-based ecosystem together with external stakeholders unfold as they address four principal areas: *interest or stakeholder groups, culture, resources, and infrastructure* (cf. Stam & Van de Ven, 2021).

Stakeholders and Culture

In this interplay between internal elements of ecosystems within HEIs, and the support from external stakeholders in the region, the active cooperation between universities and other actors will be favorable for an effectively functioning EE (Greene et al., 2010). While current universities may cooperate more intensively with external stakeholder groups than they did in the past (Galvao et al., 2020), such cooperation still needs to be managed in light of possible different, if not divergent, interests of the stakeholders to support sustainable EE and subsequent sustainable venturing (Bischoff & Volkmann, 2018). Bischoff and Volkmann (2018, p. 29) provide an overview of stakeholders involved in EE in HEIs, including (non)governmental institutions, SMEs and large businesses, investors, consultants, incubators and technology parks, student organizations, alumni, other HEIs, and entrepreneurs. Among these stakeholders, nongovernmental institutions or activist groups raising novel ecological issues, as well as the inspiration and advice of other sustainable entrepreneurs and established businesses with a sustainable mission, appear to be particularly important for sustainable entrepreneurial ecosystems (Theodoraki et al., 2018) and the functioning of EE within these systems. For example, stakeholder actors may provide ecological and social problems for students to work on in EE courses, or stakeholders can offer advice and support to student teams designing sustainable business models.

Principally, external stakeholders may be both providers of inputs to EE at universities (e.g., in roles such as lecturers, guest speakers, coaches, or judges in entrepreneurship courses, projects, or events) as well as beneficiaries of EE (e.g., in cooperation with sustainable student startsups, providing specific know-how or technology access, or as future employers of entrepreneurially minded graduates) (Bischoff et al., 2018). This balance of input supplied by external ecosystem stakeholders to improve sustainable EE and help sustainable entrepreneurs, and the outputs these external stakeholders receive from HEIs as beneficiaries of sustainable EE, may be critical. This issue has been discussed since the early days of the EXIST program. The main problem is that it is unclear whether external stakeholders in the entrepreneurship support networks created under the umbrella of the EXIST program perceive sufficient benefits to remain as active contributors to university entrepreneurship in the long run (cf. Grünhagen et al., 2005). The long-term contributions of external stakeholders will be significant for the longtime horizon required for sustainable entrepreneurship. For example, external businesses or investors may find it harder to see sufficient benefits and rewards from providing resources to sustainable venture projects because of the longtime horizon of investments and the challenge of dispersed benefits of such ventures. For the composition of sustainable ecosystems around HEIs, this may imply being even more selective in organizing cooperation with stakeholders with a dedicated sustainable mission. They would support sustainable campus ventures mainly because of their societal benefits, which means they serve as impact investors rather than providing resources for financial returns.

Further, this delicate balance of contributions and benefits for stakeholders of sustainable entrepreneurship ecosystems around HEIs depends on how much effort HEIs put into sustainable EE and the support of individual university spinoff creations. In this context, the EXIST program's fourth phase focused on fostering entrepreneurial universities (*Gründerhochschulen*) (see Sect. 3.1). This phase was dedicated to promoting the efforts of those HEIs that truly wanted to increase university-wide entrepreneurship offerings alongside specific themes such as technology, social, or sustainable entrepreneurship in an attempt to involve faculties other than business and economics, such as the natural sciences, arts and humanities, or architecture and design. However, it is still somewhat unclear if an external policy program can induce a culture of entrepreneurship into what has traditionally been a different academic culture within German and other university institutions (Volkmann & Grünhagen, 2014). While a transformative change toward a fully fledged entrepreneurship culture may be more difficult to achieve, the German higher education sector has made visible progress in recent years—at least in numbers (e.g., the number of EE courses or universities with entrepreneurship professorships) to improve the breadth and depth of EE and startup support (Funke & Schröder, 2020; Sternberg et al., 2021). To continue this trend requires the expansion of a sustained resource base for entrepreneurship in German HEIs.

Resources and Infrastructure

A functioning and generous provision of EE at German HEIs seems to depend significantly on a sufficient pool of financial and human capital *resources* (in particular, entrepreneurship lecturing and startup coaching personnel) (Funke & Schröder, 2020; also cf. Mukesh & Pillai, 2020). However, entrepreneurship at German HEIs is still predominantly driven by temporary third-party funding and temporary employment (Funke & Schröder, 2020). This may result in an increased fluctuation of personnel when entrepreneurship projects end or when temporary employment, for example that of junior faculty involved in entrepreneurship teaching is terminated. In turn, this may lead to a withdrawal of staff hours spent on entrepreneurship activities (such as preparing and delivering EE courses or extracurricular activities) as newly employed personnel has to focus on building their academic careers by doing more research and publication work rather than entrepreneurship teaching. With regard to the latter, Funke and Schröder (2020, p. 24) further assert that university members often prioritize academic reputation (e.g., through research publications), while engagement in practical entrepreneurship support and EE activities come second (also cf. EU, 2015, for a discussion of incentive structures for EE in academia in the European context). Correspondingly, active engagement in curricular and extracurricular entrepreneurship—including teaching elaborate, interactive EE courses and providing coaching to student startup teams—depends more on the intrinsic motivation of individual university employees rather than being institutionalized and embedded in university-level strategy. Only a small number of universities in Germany have implemented additional incentives such as financial allowances for engagement in entrepreneurship support or relief compensations in academic duties for engaging in academic startup projects. Accordingly, Funke and Schröder (2020) assert that there is no common standard of incentive structures for entrepreneurship at German HEIs yet, leaving ample room for improvement.

In the absence of sufficient institutional incentives for entrepreneurship, there may be opportunities for further sustainable EE in HEIs as engagement in the environment and climate protection is growing among academics and young students. There is an increased interest in contributing to sustainable entrepreneurial ecosystems (Theodoraki et al., 2018; cf. Wagner et al., 2021). Institutionally, the verdict remains that a stable perpetuation of financial and personnel resources in entrepreneurship at German HEI is lacking, with universities demanding more continuous funding from the government (Funke & Schröder, 2020).

Beyond simply increasing funding for EE and overall university entrepreneurship in the German HEI sector, further institutional stabilization will be key. This would also make it easier for entrepreneurship advocates at German HEIs to convince university management to prop up and ultimately expanding existing resource bases. For example, a further buildup of such permanent *infrastructure* involves establishing university-wide entrepreneurship centers, scientific incubators, coworking spaces, etc. German HEIs typically organize startup and legal advice consultation, mentoring, meetings with potential investors through organizational units such as entrepreneurship centers, incubators, or startup offices (Bundesministerium für Wirtschaft & Energie, 2020; Funke & Schröder, 2020; Sternberg et al., 2021). This general infrastructure may also be utilized for supporting sustainable entrepreneurship with many scientific ideas. Potential uses of novel technologies developed at universities may grow into sustainable innovation and sustainable business models. In addition, more specific infrastructure to support sustainable entrepreneurship can be established. Tiemann et al. (2018) provide a case study of a German "humanistic, yet entrepreneurial university" and its on-campus activities for sustainable entrepreneurship. A fundamental form of institutional establishment in this regard is increased cooperation between entrepreneurship-related faculties (e.g., business and economics) and faculties or departments for sustainability or environmental science, which would intensively explore ecological challenges and provide initial technological solutions derived from research (Wagner et al., 2021). Finally, the different resources and types of infrastructure for organizing activities for sustainable EE and sustainable venturing lead to considerations about the overall strategy development for fostering sustainability and entrepreneurship within German HEIs and in their regions.

6.5 Conclusion: Nurturing Development of Sustainable EE in German HEIs

To illuminate some of the most important strategic issues in paving the way for HEIs to further develop sustainable EE and overall university entrepreneurship in Germany, the following aspects will be addressed in this final section: capacity-building and strategizing for supporting sustainable EE; institutional framing and utilization of competence frameworks relating to sustainable EE; supporting education for sustainable entrepreneurs and scaling their ventures; and open-ended long-term development of sustainable EE.

6.5.1 Strategizing and Capacity-Building for Supporting Sustainable EE

Most importantly, the varied experiences of promoting sustainable entrepreneurship at German HEIs indicate that there are different ways and configurations of university strategies and organizational structures (e.g., Wagner et al., 2021, who explore support activities for sustainable entrepreneurship at German HEIs in two different Länder states). Similarly, Tiemann et al. (2018, p. 102) "identified different design strategies for university support of sustainable entrepreneurship" in German universities. Additive strategic approaches that treat sustainability and entrepreneurship as two distinct themes within university teaching seem feasible for expanding the reach of EE to students and other target groups. However, "integrative approaches, which clearly define the concept of sustainable entrepreneurship and are focused on it are important to take advantage of the specific opportunities and challenges of sustainable entrepreneurs" (p. 106). In terms of developing a profile of sustainable EE at HEIs, Tiemann et al. considered different development strategies (top-down, bottom-up, or combined). In the first approach, university top management would craft a strategy for sustainable entrepreneurship and develop organizational structures and elements of organizational culture. But this may be difficult without the initial legitimization of sustainable entrepreneurship for university ecosystems where many science disciplines compete for resources and attention (cf. Theodoraki et al., 2018). The bottom-up approach where support activities are initiated by individual university members (e.g., lecturers teaching sustainable EE) might lack long-term institutionalization and resource support for a substantial increase in entrepreneurship and outreach across campus beyond singular sustainable EE offerings by individual university staff.

However, a combined approach may be suitable to mitigate and manage the difficulties of pure top-down or bottom-up approaches when augmented with accompanying incentive structures for staff to engage in sustainable entrepreneurship and further *capacity-building* and institutional framing (Tiemann et al., 2018). Examples of capacity-building are long-term infrastructural elements for sustainable EE such as entrepreneurship and sustainability centres, professorships, departments, or other inter- and transdisciplinary units on campus. In addition, establishing novel degree programs on sustainability management or sustainable innovation and entrepreneurship may also contribute to institutional anchoring. This is because such degree courses or curricular modules require at least medium-term teaching inputs, which HEIs must provide once they are accredited. Often, the feasibility and emergence of such novel degree programs or fields of study may depend on the forward-looking strategy of the university in terms of future fields of study and research based on trends or challenges in society in general (e.g., digitalization, climate change, or aging populations). In this regard, institutional framing may be another way to develop sustainable EE development.

6.5.2 Institutional Framing and Competence Framework of Sustainable EE

In addition to building adequate long-term capacities for supporting sustainable EE and green entrepreneurship, Tiemann et al. (2018) offer another interesting approach for the continued support of sustainable EE: the notion of *institutional framing*. Institutional frames that HEIs could adopt for sustainable entrepreneurship may be rooted in the different UN Sustainable Development Goals (Fleaca et al., 2018), as universities have been tasked with promoting sustainability principles since the 1990s (United Nations, 2015). The most potent frames may be the formation of sustainable universities (Disterheft et al., 2015) or entrepreneurial universities (originally Clark, 1998). Behind such powerful yet controversial frames for university-wide organizational blueprints, more instrumental elements of organizational strategy in HEIs may also be helpful in promoting sustainable EE and indeed other entrepreneurship themes such as social, cultural, or technology entrepreneurship. For example, Funke and Schröder (2020) reported that more than 80% of German HEIs have negotiated target agreements with state ministries regarding the promotion of entrepreneurship and the supply of EE and startup support services. Further, about 40% of HEIs have a codified transfer strategy, including the support of university spinoffs. However, while sustainability and entrepreneurship may become more relevant to German HEIs as individual themes, there is still room for improvement in terms of having an integrated approach for promoting sustainable entrepreneurship (cf. Tiemann et al., 2018). In terms of the universe of EE, the integration of adjacent competence frameworks (such as those developed within the EU) may be an interesting way forward to organize education for different types of enterprise and for building entrepreneurial mind sets for sustainable, green, social, cultural, digital, or technology entrepreneurship.

Frames or toolboxes of competences provide an orientation for educating students to master challenges in various domains of society with sustainable transformation among them as well as for lecturers and education managers. For example, the European frameworks DigComp and EntreComp, and, more recently, LifeComp and GreenComp, can be employed for sustainable EE. The GreenComp framework, launched at the start of 2022, covers sustainability for lifelong learning based on the European Council's recommendation for learning environmental sustainability (EU, 2021). In addition to individual competence frames for teaching and learning, Mets et al. (2021) suggest developing a "green transformation competence" framework based on principles of sustainable EE in an integrated transdisciplinary approach to education for sustainable development. The main idea of this multi-competence framework is to interlink the above different frameworks in an EE approach that will develop "an active, informed, responsible, yet sustainable, living ecosystem-oriented and green orientation of citizens in education systems" (p. 1). Such an enterprising "transformation literacy" of participants in sustainable entrepreneurial ecosystems—where students may become ambassadors for ecological and social issues in their university region—will be critical for not only the initial formation of sustainable venture projects, but also for the competent support of the further expansion of sustainable ventures to increase the impact of the products and service solutions. This process will require competences from the various frameworks highlighted above. For example, sustainable entrepreneurs and their supporting ecosystem stakeholders may use digital or other technologies in novel products to reduce carbon emissions or in virtual platforms contributing to the circular economy.

6.5.3 Supporting EE for Sustainable Entrepreneurs and Scaling Ventures

For good reason, HEIs' sustainable EE may be focused on developing responsible entrepreneurial mind sets in students, with some becoming nascent sustainable entrepreneurs. Ecosystem stakeholders outside of the universities may contribute to supporting these entrepreneurial activities, for example, by participating in different EE activities. In addition, to truly contribute to sustainable transformation, sustainable ventures originating from the university campus and the stakeholders they team up with in their region need to move toward the landscape or regime levels. In this scaling process of sustainable transformation emerging from initial classroom sustainable EE, entrepreneurial ecosystems in university regions are key as they helping sustainable entrepreneurs to overcome challenges by providing long-term resources, including alternative impact investments, and the initial legitimacy. The latter would convince, for example, businesses or policymakers at the landscape or regime levels (cf. DiVito & Ingen-Housz, 2021).

The overall support of sustainable entrepreneurs from internal university stakeholders and external regional stakeholders may unfold alongside many phases of a process that entails the recognition of a social or ecological issue and corresponding opportunities; the crafting of a triple-bottom-line offering that addresses the problem issue; and the follow-up market entry and expansion (Belz & Binder, 2017). In view of the complexity of the establishment process of sustainable entrepreneurship, it may well be that there will neither be a single "blueprint model" for sustainable EE, nor a uniform support infrastructure for scaling sustainable ventures from university across different segments such as clean tech, renewable energies, solutions for the circular or the sharing economy, or the preservation of biodiversity. At the same time, with regard to general support strategies and infrastructural setups for university entrepreneurship and EE, the substantial impact of policy support programs such as EXIST may have led to some organizational isomorphism of "acceptable approaches" or at least operational similarities in capacity-building. The latter includes establishing entrepreneurship chairs, centers, or incubators at HEIs (cf. Funke & Schröder, 2020).

6.5.4 Open-Ended Long-Term Development of Sustainable EE

Correspondingly, Tiemann et al. (2018, p. 106) summarize the different approaches of organizing activities for sustainable entrepreneurship in their sample of German universities: "It remains open whether support systems will converge to 'one best way' to organise university support systems in the future or whether there is simply no 'one best way' to support sustainable entrepreneurship but rather various design configurations that correlate with different contingency factors." In terms of scaling sustainable ventures or other sustainable innovations developed in the niche, Augenstein et al. (2020) recommend allowing for more contingencies as well as creating learning and discourse spaces within regional niche ecosystems to explore alternative paths to support the solution of ecological problems.

Universities and other HEIs may be in a good position with their plurality of disciplinary fields of study and research. Additionally, German HEIs' autonomous—or at least partly independent—nature may enable intensive experimentation with different strategies, structures, processes, and didactical approaches for sustainable EE and university entrepreneurship in general. However, it will also be crucial for EE in Germany's higher education sector to retain and preserve those configurations of delivering EE and supporting university spinoffs alongside different forms of entrepreneurship in the long run. Toward this end, the German higher education system and surrounding policy streams of education, research, and entrepreneurship policymaking may be in a somewhat worse position. First, experimentation with different configurations of EE and university entrepreneurship will require effective and high-quality evaluation of the EE instruments and startup support formats employed in university practice. Concerning this, Funke and Schröder (2020) report that while monitoring startup spinoffs by German HEIs is improving, only about half of HEIs evaluate their own approaches toward EE and the support of university entrepreneurship. A better evaluation may provide more effective evidence-based policymaking in EE, as indicated in a report to the European Commission on fostering EE for developing a digital, green, and resilient Europe (Lilischkis, et al., 2021). Second, as discussed throughout this chapter, entrepreneurship and EE at German HEIs depend substantially on external, fixed-term, third-party funding for their internal EE operations and activities. This policy approach risks the termination of well-functioning configurations of EE and entrepreneurship support, particularly of more "vulnerable," less institutionalized extracurricular EE activities, which often are very interactive, hands-on, and practical in cooperation with stakeholders. Generally, the short-term external funding of EE and related activities affects the initial experimentation with different setups of EE and spinoff support across organizations in German higher education, and the subsequent retention of successfully working configurations.

Interestingly, providers of third-party funding for university entrepreneurship to the German higher education sector (including governmental sources such as EXIST or the Exzellenz Start-up Center program) have demanded that public universities develop funding strategies from their own baseline budget sources for the further continuation of these third-party projects. For sustainable EE, in particular, sustainability seems to be an opportunity to become a "century issue" that will shape university strategies for a long time. A possible policy window prompted by the convergence of policy streams in environmental, education, and entrepreneurship may increase the impact of sustainable EE within HEIs supported by external policymaking (e.g., a sustained provision of long-term funding for sustainable entrepreneurship and other themes) and internal obligations of HEIs for a growing—and possibly permanent—engagement in education for sustainable development.

References

Audretsch, D. B., & Belitski, M. (2016). Entrepreneurial ecosystems in cities: Establishing the framework conditions. *Journal of Technology Transfer, 42*(5), 1030–1051.

Augenstein, K., Bachmann, B., Egermann, M., Hermelingmeier, V., Hilger, A., Jaeger-Erben, M., & Von Wirth, T. (2020). From niche to mainstream: The dilemmas of scaling up sustainable alternatives. *GAIA-Ecological Perspectives for Science and Society, 29*(3), 143–147.

Béland, D., & Howlett, M. (2016). The role and impact of the multiple-streams approach in comparative policy analysis. *Journal of Comparative Policy Analysis: Research and Practice, 18*(3), 221–227.

Belz, F. M., & Binder, J. K. (2017). Sustainable entrepreneurship: A convergent process model. *Business Strategy and the Environment, 26*(1), 1–17.

Bischoff, K. (2021). A study on the perceived strength of sustainable entrepreneurial ecosystems on the dimensions of stakeholder theory and culture. *Journal of Small Business Economics, 56*, https://doi.org/10.1007/s11187-019-00257-3

Bischoff, K., & Volkmann, C. K. (2018). Stakeholder support for sustainable entrepreneurship—A framework of sustainable entrepreneurial ecosystems. *International Journal of Entrepreneurial Venturing, 10*(2), 172–201.

Bischoff, K., Volkmann, C. K., & Audretsch, D. B. (2018). Stakeholder collaboration in entrepreneurship education: An analysis of the entrepreneurial ecosystems of European higher educational institutions. *Journal of Technology Transfer, 43*(1), 20–46.

Blignaut (Taljaard), A., & Botha, M. (2022). A conceptual competencies-based framework. to enhance the innovation capacity of fourth industrial revolution entrepreneurs. *The International Journal of Entrepreneurship and Innovation, 0*(0). https://doi.org/10.1177/14657503221115883

Bundesministerium für Wirtschaft und Energie. (2020). *Öffentliche Infrastruktur in Deutschland: Probleme und Reformbedarf*. Gutachten des Wissenschaftlichen Beiratsbeim Bundesministerium für Wirtschaft und Energie, Berlin, Germany: Bundesministerium für Wirtschaft und Energie. [In German].

Chalmers, D., et al. (2021). Artificial intelligence and entrepreneurship: Implications for venture creation in the fourth industrial revolution. *Entrepreneurship Theory and Practice, 45*(5), 1028–1053.

Clark, B. (1998). *Creating entrepreneurial universities: Organizational pathways of transformation.* Pergamon.

Disterheft, A., Caeiro, S., Azeiteiro, U. M., & Filho, W. L. (2015). Sustainable universities—A study of critical success factors for participatory approaches. *Journal of Cleaner Production, 106*, 11–21.

DiVito, L., & Ingen-Housz, Z. (2021). From individual sustainability orientations to collective sustainability innovation and sustainable entrepreneurial ecosystems. *Small Business Economics, 56*(3), 1057–1072.

European Union. (2014). *Green action plan for SMEs: Enabling SMEs to turn environmental challenges into business opportunities.*

European Union. (2015). *Supporting the entrepreneurial potential of higher education* (EU Commission Report). Publications Office of the European Union.

European Union. (2021). *The impact of covid-19 on higher education: A review of emerging evidence.* Publications Office of the European Union.

Fleaca, E., et al. (2018). Aligning Strategy with Sustainable Development Goals (SDGs): Process scoping diagram for entrepreneurial Higher Education Institutions (HEIs). *Sustainability, 10*, 1032–1047.

Funke, A., & Schröder, E. (2020). *Gründungsradar 2020: Wie HochschulenUnternehmensgründungenfördern.* Stifterverband für die Deutsche Wissenschafte.V. Stifterverband. [In German].

Galvão, A. R., Marques, C. S. E., Ferreira, J. J., & Braga, V. (2020). Stakeholders' role in entrepreneurship education and training programmes with impact on regional development. *Journal of Rural Studies, 74*, 169–179.

Gast, J., Gundolf, K., & Cesinger, B. (2017). Doing business in a green way: A systematic review of the ecological sustainability entrepreneurship literature and future research directions. *Journal of Cleaner Production, 147*, 44–56.

Geels, F. W. (2011). The multi-level perspective on sustainability transitions: Responses to seven criticisms. *Environmental Innovation and Societal Transitions, 1*(1), 24–40.

Greene, P. G., Rice, M. P., & Fetters, M. L. (2010). University-based entrepreneurship ecosystems: Framing the discussion. In M. L. Fetters, P. G. Greene, M. P. Rice, & J. S. Butler (Eds.), *The development of university-based entrepreneurship ecosystems: Global practices* (pp. 1–15). Edward Elgar.

Grünhagen, M., Koch, L. T., & Saßmannshausen, S. P. (2005). Kooperation in EXIST-Gründungsförderungsnetzwerken—Eine explorative Untersuchung zur Bedeutung von Promotorenfunktionen. In A. K. Achleitner, H. Klandt, L. T. Koch, & K. I. Voigt (Eds.), *Jahrbuch Entrepreneurship 2004/05* (pp. 319–338). Springer. [In German].

Hahn, T., Pinkse, J., Preuss, L., & Figge, F. (2015). Tensions in corporate sustainability: Towards an integrative framework. *Journal of Business Ethics, 127*(2), 297–316.

Hochschulrektorenkonferenz. (2017). *Organisationsstrukturen an Hochschulen.* Retrieved from https://www.hrk.de/positionen/beschluss/detail/organisationsstrukturen-an-hochschulen/. [In German].

Hochschulrektorenkonferenz. (2021). *Hochschulautonomie.* Retrieved from https://www.hrk.de/themen/hochschulsystem/hochschulautonomie/. [In German].

Högsdal, N., Münch, J., & Bauer, H. (2020). *Entrepreneurship meets Education: Förderung der Gründungskultur an Hochschulen: Ergebnisse.* Erfahrungen und Erfolgsfaktorenaus dem Verbundprojekt Spinnovation. [In German].

Hörisch, J. (2015). The role of sustainable entrepreneurship in sustainability transitions: A conceptual synthesis against the background of the multi-level perspective. *Administrative Science, 5*(4), 286–300.

Kalden, J. N., Cunningham, J., & Anderson, A. R. (2017). The social status of entrepreneurs: Contrasting German perspectives. *International Journal of Entrepreneurship and Innovation, 18*(2), 91–104.

Kaminski, H. (2016). Entrepreneurship Education und ökonomische Bildung—Fehlerkultur und Laiendenken als Hemmfaktoren für Gründungsaktivitäten. In B. Greimel-Fuhrmann & R. Fortmüller (Eds.), *Facetten der Entrepreneurship Education: Festschrift für Josef Aff* (pp. 27–35). MANZ Verlag. [In German].

Kansheba, J. M. P., & Wald, A. E. (2020). Entrepreneurial ecosystems: A systematic literature review and research agenda. *Journal of Small Business and Enterprise Development, 27*(6), 943–964.

Kingdon, J. W. (1984). *Agendas, alternatives and public policies.* Little, Brown and Company.

Klusmeyer, J., Schlömer, T., & Stock, M. (2015). Entrepreneurship Education in der Hochschule. *Zeitschrift Für Hochschulentwicklung, 10*(3), 9–22.

Kuckertz, A., Berger, E. S., & Prochotta, A. (2020). Misperception of entrepreneurship and its consequences for the perception of entrepreneurial failure—The German case. *International Journal of Entrepreneurial Behavior and Research, 26*(8), 1865–1885.

Kulicke, M. (2014). *15 Jahre EXIST "Existenzgründungenaus der Wissenschaft": Entwicklung des Förderprogramms von 1998 bis 2013.* Karlsruhe. [In German].

Kulicke, M. (2018). *EXIST-Gründungskultur—Die Gründerhochschule: Erste Ergebnisse zu Förderwirkungen und Weiterführung der Unterstützungsmaßnahmen nach Förderende.* Karlsruhe. [In German].

Kuratko, D. F., Fisher, G., Bloodgood, J. M., & Hornsby, J. S. (2017). The paradox of new venture legitimation within an entrepreneurial ecosystem. *Small Business Economics, 49*(1), 119–140.

Lackeus, M. (2015). *Entrepreneurship in education: What, why, when, how.* OECD.

Liguori, E., & Winkler, C. (2020). From offline to online: Challenges and opportunities for entrepreneurship education following the COVID-19 pandemic. *Entrepreneurship Education and Pedagogy, 3*(4), 346–351.

Lilischkis, S., et al. (2021). *A guide to fostering entrepreneurship education. Five key actions towards a digital, green and resilient Europe* (Report to the European Commission).

Loorbach, D., & Wijsman, K. (2013). Business transition management: Exploring a new role for business in sustainability transitions. *Journal of Cleaner Production, 45*, 20–28.

Lüdeke-Freund, F. (2020). Sustainable entrepreneurship, innovation, and business models: Integrative framework and propositions for future research. *Business Strategy and the Environment, 29*(2), 665–681.

Mets, T., et al. (2021). Entrepreneurship Education – Challenges for Green Transformation. *Administrative Sciences, 11*, 15–29.

Metzger, G. (2021). *KfW-Gründungsmonitor 2021—Gründungstätigkeit 2020 mit Licht und Schatten: Corona-Krise bringt Tiefpunkt im Vollerwerb, birgt für viele aber auch Chancen.* Frankfurt. [In German].

Moberg, K., Stenberg, E., & Vestergaard, L. (2015). *Impact of Entrepreneurship Education in Denmark*. Danish Foundation for Entrepreneurship.

Mukesh, H. V., Pillai, K. R. (2020). Role of institutional ecosystem in entrepreneurship education: An empirical reiteration. *Journal of Entrepreneurship, 29*(1), 179–205.

O'Shea, G., Farny, S., & Hakala, H. (2021). The buzz before business: A design science study of a sustainable entrepreneurial ecosystem. *Small Business Economics, 56*(3), 1097–1120.

OECD. (2018). *The future of education and skills: Education 2030*. OECD.

OECD. (2020). *Bildung auf einen Blick*. Retrieved from https://www.oecd.org/berlin/publikationen/EAG2020_CN_DEU_de_FINAL_FINAL.pdf. [In German].

Pankov, S., Velamuri, V. K., & Schneckenberg, D. (2021). Towards sustainable entrepreneurial ecosystems: Examining the effect of contextual factors on sustainable entrepreneurial activities in the sharing economy. *Small Business Economics, 56*(3), 1073–1095.

Patzelt, H., & Shepherd, D. A. (2011). Recognizing opportunities for sustainable development. *Entrepreneurship Theory & Practice, 35*(4), 631–652.

Regele, M. D., & Neck, H. M. (2012). The entrepreneurship education sub-ecosystem in the United States: Opportunities to increase entrepreneurial activity. *Journal of Business and Entrepreneurship, 23*(2), 25–47.

Ribeiro, A. T. V. B., Ferragi, C. A., Trivinho-Strixino, F., & Cardoso, A. C. F. C. (2020). Entrepreneurship education going remote: A response to Covid-19 restrictions. *Journal of Entrepreneurship Education, 23*(2), 1–9.

Schaltegger, S., & Wagner, M. (2011). Sustainable entrepreneurship and sustainability innovation: Categories and interactions. *Business Strategy and the Environment, 20*(4), 222–237.

Sedlacek, S. (2013). The role of universities in fostering sustainable development at the regional level. *Journal of Cleaner Production, 48*, 74–84.

Stam, E., & Van de Ven, A. (2021). Entrepreneurial ecosystem elements. *Small Business Economics, 56*(2), 809–832.

Sternberg, R., Gorynia-Pfeffer, N., Stolz, L., Schauer, J., Baharian, A., & Wallisch, M. (2021). *Global entrepreneurship monitor: UnternehmensgründungenimweltweitenVergleich*. RKW Kompetenzzentrum. [In German].

Theodoraki, C., Messeghem, K., & Rice, M. P. (2018). A social capital approach to the development of sustainable entrepreneurial ecosystems: An explorative study. *Small Business Economics, 51*(1), 153–170.

Tiemann, I., Fichter, K., & Geier, J. (2018). University support systems for sustainable entrepreneurship: Insights from explorative case studies. *International Journal of Entrepreneurial Venturing, 10*(1), 83–110.

United Nations. (2015). *Transforming our world: The 2030 agenda for sustainable development*. Retrieved from https://www.unfpa.org/sites/default/files/resource-pdf/Resolution_A_RES_70_1_EN.pdf.

Volkmann, C. K., & Audretsch, D. B. (2017). Introduction: The mandate for Entrepreneurship Education. In C. K. Volkmann & D. B. Audretsch (Eds.), *Entrepreneurship education at universities: Learning from twenty European cases* (pp. 1–12). Springer.

Volkmann, C., & Grünhagen, M. (2014). Integrated support for university entrepreneurship from entrepreneurial intent towards behaviour: The case of the German "EXIST" policy programme. In A. Fayolle & D. T. Redford (Eds.), *Handbook on the Entrepreneurial University* (pp. 225–247). Edward Elgar.

Volkmann, C., Fichter, K., Klofsten, M., & Audretsch, D. B. (2021). Sustainable entrepreneurial ecosystems: An emerging field of research. *Small Business Economics, 56*(3), 1047–1055.

Volkmann, C. (2009). *Educating the wave of entrepreneurs* (Report to the World Economic Forum).

Wagner, M., Schaltegger, S., Hansen, E. G., & Fichter, K. (2021). University-linked programmes for sustainable entrepreneurship and regional development: How and with what impact? *Small Business Economics, 56*(3), 1141–1158.

Wells, P. (2016). Economies of scale versus small is beautiful: A business model approach based on architecture, principles and components in the beer industry. *Organization & Environment, 29*(1), 36–52.

Welter, F., Baker, T., & Wirsching, K. (2019). Three waves and counting: The rising tide of contextualization in entrepreneurship research. *Small Business Economics, 52*(2), 319–330.

Wissema, J. G. (2009). *Towards the third generation university: Managing the university in transition.* Edward Elgar Publishing.

York, J. G., & Venkataraman, S. (2010). The entrepreneur–environment nexus: Uncertainty, innovation, and allocation. *Journal of Business Venturing, 25*(5), 449–463.

Chapter 7
Entrepreneurship Education in the Republic of Korea

Tae Uk Ahn

7.1 Introduction

In today's era of hyper-convergence, hyper-connectivity, and hyper-intelligence, Korean society requires considerable changes in its university education system. As rapid transformations in the external environment lead to doubts about the existing knowledge-centered education, entrepreneurship education (EE) emerges as a solution for the current education system as well as economic and societal problems (e.g., slow economic growth and youth unemployment). Academia proposes that EE can guide the way education should move forward in terms of educational philosophy, university system, curriculum, and creative talents for the Fourth Industrial Revolution.

Implementing changes requires several preconditions. First, it is necessary to change the curriculum authority system which is aligned with the university education system. This implies an overhaul of the education system and changes in the administrative process that support the university education system, thereby reducing its administrative bureaucracy. Second, the reorganization of the university curriculum should be emphasized as a way to build a desirable future competency. Third, particularly for future talent, it is necessary to improve the competency-oriented curriculum required for the future society through a "change of the future talent figure and education system to lead the intelligent information society".

Recently, the Korean society has begun to recognize the importance of EE. The government is paying attention to the revitalization of startups through EE (Ministry of Economy and Finance, 2019). Entrepreneurship is attracting attention as an alternative education to adapt to new changes especially when there have been significant changes in the external environment and hence increased economic uncertainty.

T. U. Ahn (✉)
Institute for Startup KAIST (Korea Advanced Institute of Science and Technology), Daejeon, Republic of Korea
e-mail: helloceo777@gmail.com

X. Xu, *Comparative Entrepreneurship Education*,
https://doi.org/10.1007/978-981-99-1835-5_7

Entrepreneurship can change the social paradigm through creative destruction as well as innovation, and drive social change. Entrepreneurial activities are important driving forces for economic growth and for creating high-quality jobs. There are tremendous benefits for supporting startups in view of the technology possessed by the university in the entrepreneurial ecosystem. However, EE is not just for encouraging startup attempts. Although the basic purpose of EE is the commercialization of business, it also cultivates the mindset, attitude, and behavior for an individual's career development. It aims to nurture young innovative talents by fostering collaboration, creativity, and problem-solving skills based on the spirit of entrepreneurship. Therefore, universities can be a place for cultivating entrepreneurship and supporting EE, regardless whether students start their own business or not. Ahn and Park (2018) argue that EE has a positive and significant impact on the career preparation of college students.

This chapter takes a detailed look at the current status, characteristics, case studies, and future directions of EE in South Korea.

7.2 Historical Development of EE in the Republic of Korea

EE in Korea began as a field of business administration in the late 1970s. The main subject was "Small Business Management," which focused on cultivating business consulting capabilities centered on business administration, and it began to grow in the late 1990s. Interest in domestic startups began to increase with the enactment of the Small and Medium Business Startup Support Act (Law No. 3831) on May 12, 1986. After that, the government began promoting the emergence of small and medium-size enterprises (SMEs) by introducing a startup support program, facilitating university-level EE, and simplifying startup registration procedures. Notably, entrepreneurship was introduced as a way to support the management of SMEs.

EE in universities began in 1983 when Ajou University's Business Administration offered the course "Small and Medium Business Theory." Since the enactment of the Small and Medium Business Startup Support Act, interest in domestic startups and publication of startup-related books increased (Han, 2007). With the support of the Small and Medium Business Administration, the introduction of Bizcool in vocational high schools in the late 1990s and the implementation of EE in nonprofit organizations began in earnest. Then, in 1992, 28 universities established entrepreneurship courses (Lee & Hwang, 2015).

In 1995, Soongsil University opened Korea's first department of SMEs. In 1996, the Small and Medium Business Administration was replaced with the Ministry of SMEs and Startups, and thereafter the Act on Special Measures for Supporting Small Businesses and Act on Special Measures for the Promotion of Venture Businesses were enacted, and startup support programs spread (Lee et al., 2013).

In Korea, starting in the 2000s, universities offered entrepreneurship as an independent regular education course, not as a part of business administration. The curriculum and contents have expanded and developed into practice-based programs.

There are 160 universities that provide entrepreneurship courses as part of the undergraduate curriculum and most of the courses combine entrepreneurship practice and management theory. Since the 2000s, EE in universities has made rapid progress, but it has been limited to general liberal arts courses related to entrepreneurship, such as "(Venture) Entrepreneurship Theory" and "Startup and Management" (Jung, 2013).

Universities also provide specific practice-based education such as domestic and overseas training, coaching, and internship in the graduate school-level curriculum. In 2004, the first graduate school in entrepreneurship was established in Korea. Five graduate schools—Chung-Ang University, Hoseo University, Hanbat University, Jinju University of Technology, and Yewon University of the Arts—started an entrepreneurship master's program. In 2007, Chung-Ang University in Seoul began to offering a PhD in entrepreneurship—the first doctoral program in Korea. In 2010, Sookmyung Women's University launched an entrepreneurship program for undergraduates, and two years later, Hoseo University founded the Faculty of Entrepreneurship (Lee & Hwang, 2015).

In March 2017, Kunsan National University founded the Department of Entrepreneurship for undergraduates—the first in a national university. In 2017, the Korean government elevated the status of the Small and Medium Business Administration to the Ministry of SMEs and Startups in order to revitalize technology startups. Currently, college entrepreneurship-related departments are popular. In particular, EE has expanded significantly with the number of enrollment for startup (entrepreneurship)-related majors rapidly increasing.

In 2017, there were 18 startup (entrepreneurship)-related university departments, with 1,074 students enrolled—2.3 and 1.9 times increase respectively compared to figures in 2015. In the case of entrepreneurship majors, there were 57 in 2017—1.8 times increase compared to 2015. In the era of the Fourth Industrial Revolution, since 2020, the demand for changes in the university curriculum has led to the continuous expansion of entrepreneurship-related majors and departments.

In South Korea, the number of entrepreneurship courses in universities has increased exponentially, focusing on cultural studies, majors, and practice-based courses. A university entrepreneurship course is a regular curriculum with credits in order to students' startup and entrepreneurship competency (Kwon, 2017) (Fig. 7.1).

From EE offerings in 28 domestic universities in 1992, this figure has grown to 4,876 in 2020. It is expected that the figure will continue to increase as EE is integrated in various ways with lifelong education.

Fig. 7.1 Entrepreneurship courses in universities. 1992–2020. *Source* Based on Lee and Hwang (2015) and Song and Kim (2020)

	Entrepreneurship courses establishment university	
	The number of Colleges with Entrepreneurship courses	Total number of Entrepreneurship courses
1992	28	•
2005	75	131
2011	140	1,913
2012	152	2,214
2013	160	3,924
	Skip	
2018	203	3,190
2019	205	4,323
2020	204	4,876

7.3 Entrepreneurship Issues in Korean Society

7.3.1 The Issue of Unemployment Among University Graduates

In South Korea, large companies in major industries have led to the economic growth. However, as the country entered a period of stagnated economic growth after the financial crisis, it is witnessing serious social issues pertaining to job creation and distribution. In addition, with the Covid-19 pandemic, uncertainty has increased. Changes in the external environment have led to a sharp decline in the demand in labor and new jobs. Therefore, South Korea is facing a serious unemployment crisis with "jobless growth." In particular, youth unemployment is a very serious problem. The youth unemployment rate is more than nine percent on average (9.8% in 2016, 9.8% in 2017, 9.5% in 2018, 8.9% in 2019, and 9.0% in 2020; Statistics Office of the Republic of Korea, 2020). Accordingly, the South Korean government has made supporting youth startups an important national task. Startups contribute to the growth of the national economy and creates jobs while increasing the economic activity of a country. According to the 2020 governmental Startup Support Company History and Performance Survey, a total of 92,367 startups received governmental support in 2019, and each company employed an average of 7.26 people. The data supports the argument that by promoting entrepreneurship, startups contribute to national economic activity and growth through job creation. Therefore, EE is an important policy agenda that can help overcome various issues, especially those affecting the younger generation.

In order to help university students launch their career after graduation and to solve the youth unemployment rate, the education curriculum requires changes. In particular, the curriculum needs to focus on cultivating innovative capabilities based on creative talents and convergence thinking. Rapid changes in the job market also demand structural changes in the university education system, for example, by introducing innovative curriculum.

7.3.2 Dynamics of Entrepreneurship in South Korea

Since the Korean War, South Korea has experienced economic miracles through fierce entrepreneurship. Peter Drucker, one of the most respected scholars in business field noted that, In his book 'The Next Society' published in 1996, he picked Korea as the country with the highest entrepreneurial spirit. In 2021, it was ranked the 10th-largest[1] economy in the world, and achieved very rapid innovation. The country was praised for its "dynamics" and "challenge spirit" that helped achieve a "miracle on the Han River."

As of 2020, the number of venture companies in South Korea has increased to 39,511, a record high for the country. In 2018, the number of venture companies with a revenue of KRW100 billion reached 587. However, the Ministry of SMEs and Startups reported that the five-year survival rate of startups is 29.2%, meaning seven out of 10 companies do not last beyond five years, which is far below the OECD average of 41.7% (KOREA Venture Business Association, 2020).

7.3.3 Reasons for Decline in Entrepreneurship

The period when entrepreneurship was at its strongest was in the 1950s and mid-1980s. Since 1980, the Korean economy had grown significantly, but entrepreneurship began to decline. In particular, the younger generation sought stability rather than challenges. Due to the rising youth unemployment rate, the global economic crisis, and the COVID-19 pandemic, society has become unstable and unpredictable. At some point, South Korean youths desired civil service jobs, and growing numbers of young people committed to civil service exams in their freshman year at university. The number of youths who applied to large companies also decreased. Hankyung Business conducted a survey of 600 South Koreans aged between 10 to over 60s on how they perceived entrepreneurship (Korean Economy BUSINESS, 2019). The results indicate a "negative perception of entrepreneurs," "preference for a stable job," and "education focused on entrance exams" as reasons for the decline in entrepreneurship. In addition, a significant number of respondents provided other reasons such as regulations, fear of failure, and strong unions. The results of this survey offer many implications, especially the need to increase the entrepreneurial spirit.

---

[1] CNBC of the United States analyzed the top 10 economies by comparing the nominal gross domestic product of each country based on the global economic forecast of the International Monetary Fund.

7.3.4 Revitalising Entrepreneurship in South Korea

In South Korea, entrepreneurship and EE have developed rapidly over the past 10 years as a new educational paradigm (Ban et al., 2008; Lee & Hwang, 2015). However, there is room for improvement.

According to the 2016–2017 Global Entrepreneurship Monitor[2] (GEM) survey based on 64 countries, EE at the level of university and lifelong education in Korea is ranked in the middle (15th). About five percent of 1,075 students considered starting a business as a career after graduation. Hence, with increased EE, individuals may be more motivated to start a business, which would enable the discovery and development of new technologies, thereby promoting innovation. In addition, EE not only nurtures creative talent, but also has a positive effect on the success of startups. From a macro perspective, EE plays an important role in economic growth throughout the country via the transfer of knowledge relating to entrepreneurship and the nurturing of talent.

In recent years, entrepreneurship in South Korea has been on the rise again. In the 2021 GEM report, South Korea rose to the ninth place out of 44 countries, up six places from the previous year. In the report, Korea's entrepreneurship ranked first in product and market dynamics; fifth for appropriateness of government startup policy; 43rd, or second last, for low fear of startup failure; and seventh for the social awareness of entrepreneurs. Preference for starting a business as a job improved to 28th from the 38th place. In addition, it is encouraging that the number of startup activities among young people (18–34 years old) also increased (17.7% in 2019 and 19.5% in 2020). These reflect the collective efforts of the Korean government, local governments, universities, and the private sector which continuously encourage entrepreneurship and EE (Fig. 7.2).

7.3.5 Depopulation and EE in South Korea

South Korea's the low birth rate and aging population are serious social issues. The prolonged low fertility rate reduces the school-age population, and which in turn affects preschool education, business incubators, teacher supply and demand, as well as the need for relocation of school facilities and university restructuring.

These issues have a ripple effect that can transform society (Kim et al., 2017). In particular, higher education is more sensitive to environmental changes because it is oriented toward specialized education. A low fertility rate amid the advent of an intelligent information society is expected to affect the higher education system among many other changes. According to the OECD, South Korea's potential growth rate decreased from five percent in the early 2000s to 2.8% in 2016, while the 2020 Korea Economic Report stated that the growing aging population and shrinking

[2] Every year, the Global Entrepreneurship Research Association publishes national rankings after collecting data from participating countries on the overall startup ecosystem.

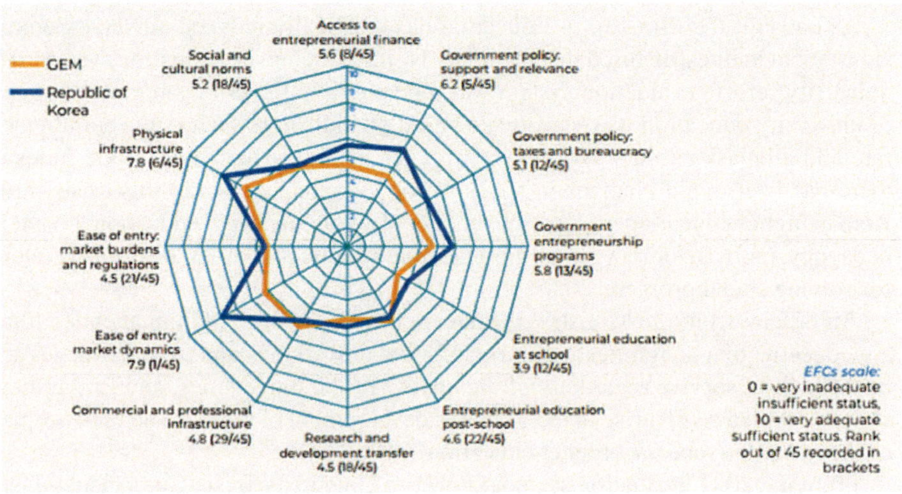

Fig. 7.2 Republic of Korea's score on entrepreneurship environment in 2020/2021 Global Entrepreneurship Monitor survey

working population would result in an average 1.2% growth rate for 2020–2060, lower than the previous average of three percent in 2005–2020. South Korea's higher education system has already begun reforming due to the low birth rate and changes in the demographic structure. In recent years, universities have been unable to fill admissions quota due to the shrinking school-age population (Fig. 7.3).

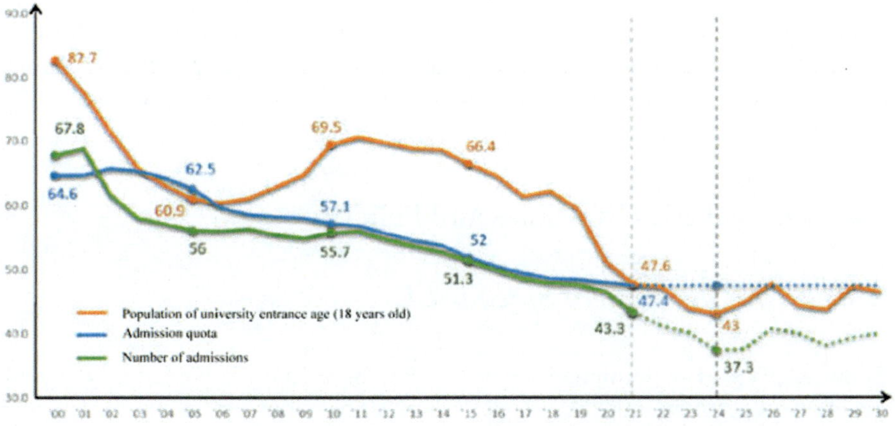

Fig. 7.3 University admissions, quota, and population of university entrance age. *Source* The number of admissions is actual students until 2021, and estimates after 2022 (Ministry of Education); *Note* *Green: Number of admissions; **Blue: Admission quota; ***Orange: Population of university entrance age (18 years old)

As a result, the Ministry of Education has drastically reduced university admissions quota and restructured the university by introducing an evaluation system. The university reform evaluation system was a response to the criticism that the quality of university education was inadequate based on indicators such as the employment rate and admissions rate. With the emergence of a new paradigm in the industry, there has been rapid changes in terms of job competency. Amid this crisis in the Korean higher education system, universities struggle to design educational content, pedagogy, innovate or reorganize the academic structure, in order to nurturing talents for solving social problems.

An official from the Ministry of Education said, "If this is left unattended, there are concerns that a significant number of local universities and junior colleges will be difficult to survive regardless of the quality of the universities, and furthermore, serious problems will arise in the balanced development of regions and enhancement of the competitiveness of higher education" (Yu, 2022).

From the 2021 evaluation system known as University Basic Competency Diagnosis Assessment, the Ministry of Education excluded 52 universities from funding for three years because they ranked poorly in the categories of self-innovation and capacity-building. A total of 233 schools including 136 general universities and 97 junior colleges receive an average of KRW4.8 billion of financial support from the Ministry of Education. As of September 18, 2021, there are 336 higher education institutions (HEIs) in Korea: 190 universities, 134 junior colleges, 10 education colleges, and two industrial colleges.

Now, the role of universities is changing from providing employment-oriented education to becoming an entrepreneurial university capable of nurturing creative talents, convergence thinking, and cultivating an entrepreneurial spirit which are required to meet social needs. In the future, students will evaluate the university, instead of universities selecting students. Students will choose universities with a more systematic educational curriculum, content, capacity-building, educational infrastructure, and career development plan. As the school-age population continues to decline, it has become very important for universities to be competitive.

7.4 Current EE Guidelines and Policies

7.4.1 The Government Spreads EE

South Korea's EE environment features institutions that are guided by the central government's policy, not the private sector. Universities are at the center of EE. Policies for nurturing talent through innovation in the higher education sector and the vocational and lifelong education system have become important. Entrepreneurship at universities is a useful means to distribute the research and technological achievements of universities to society; beyond the development of innovative technologies and products, it is possible to create a new industry, and to contribute to the development of local and national economies.

Lerner's (2000) research on government support shows that there is a difference in the growth rates of companies that receive government support and SMEs that do not receive the same support. Along with the central government's startup revitalization policy introduced in 2010, South Korea has established a leading university focused on startups, the Leaders in Industry-University Cooperation (LINC), the socially tailored Leaders in Industry-University Cooperation+ (LINC+), incubation centers, and university funds to support startups.

With the government policy, "Five-Year University Start-up Education Plan (2013–2017)," universities have sought to expand EE, increase the number of teachers specializing in the field, and create an environment for students to be involved in startups. Since 2000, EE has become an independent regular curriculum that includes practical training such as the creation and evaluation of business ideas, technical factors, and business plans. Universities have been implementing measures that foster and encourage startups, such as including business startup-related courses in the credit curriculum, and organizing startup club activities and competitions.

The "College Entrepreneurship Education Five-year Plan (2018–2022)" emphasized EE to adapt to changes in the new era (Kim & Yang, 2018). In order to create an environment where any university student can start a business, a "startup-friendly" academic system was implemented, focusing on practice-based EE rather than theory. In addition, startup credits can be exchanged between universities, and systematic support is provided to nurture outstanding startup talents through the Hope Ladder Scholarship and university startup projects (Yoon, 2018).

The South Korean government has included an "startup affinity indicator" in the selection and evaluation of financial support disbursed to universities. With the active support and will of the government, college student entrepreneurship has risen, and the number of universities providing support from education to actual startups is increasing. Infrastructure such as startup spaces, facilities, and equipment within universities is also spreading (Fig. 7.4).

7.4.2 The Value and Role of EE

Major advanced countries are actively nurturing entrepreneurial talents by promoting EE as a national task and spreading entrepreneurship in society as a whole. In order to create new jobs, it is necessary to make steady efforts at the national level to form a startup ecosystem. South Korea provides national support for youth entrepreneurship in terms of job creation and synergy effects. The spread of EE within universities is the result of a general consensus on the importance of entrepreneurship in the national economy (Birch, 1987; Matlay & Westhead, 2007; Vanevenhoven, 2013). A survey by Higher Education in Korea Service from March 2019 to February 2020 found that various entrepreneurship courses are introduced in university curriculum,

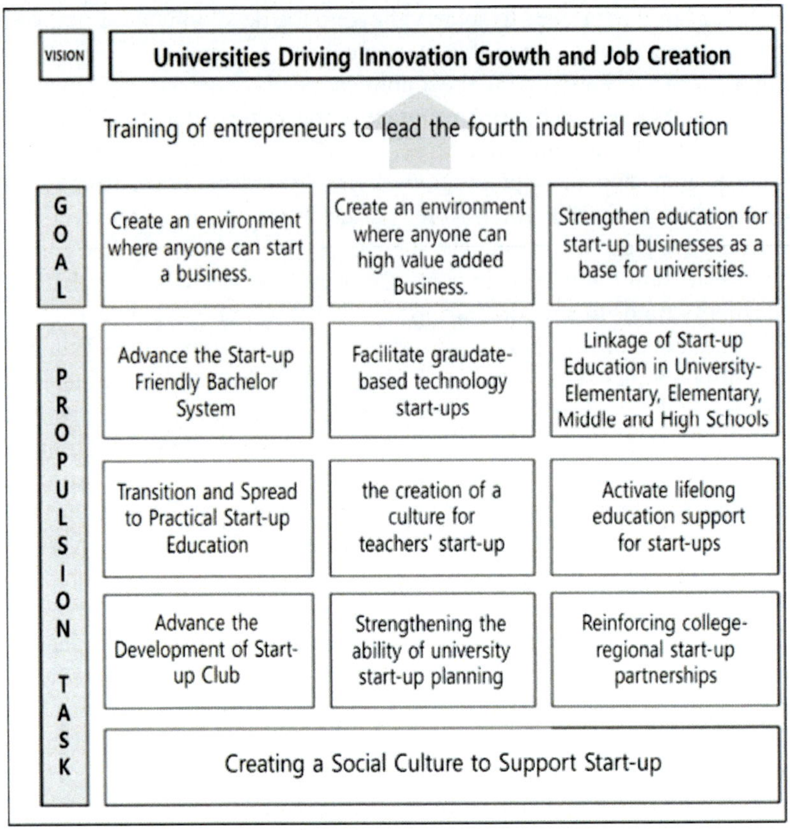

Fig. 7.4 South Korea's second five-year plan for EE in universities (2018–2022)

and not only in metropolitan universities but also major regional universities offer many entrepreneurship courses in their majors and liberal arts subjects. Among the top 15 universities receiving EE support from the government, an average of 100 entrepreneurship courses are offered per university. This is a radical change in just a few years, and it is evident that EE has been successfully integrated into the university curriculum. Despite this quantitative spread, procedures such as setting mid- to long-term goals that reflect the needs of consumers, local communities, teachers, and dedicated organizations, are still not systematically included as part of the university program development (Hong & Gang, 2016). EE in universities should not just focus on education for acquiring specific skills necessary for starting a business; it should also focus on integration into the regular curriculum so that entrepreneurship can be cultivated in all academic fields (Lee & Park, 2014).

Entrepreneurship courses are mostly theory-oriented, and the number of practice-based EE courses is still insufficient, accounting for only 30%. Accordingly, a university startup infrastructure survey in 2018 shows that the number of startups per domestic university was 9.7, the number of founders among students who completed a startup course was 0.41%, and the number of founders among students participating in a startup club was 3.71% (KISED, 2019). The results indicate that There is a lack of curriculum to learn entrepreneurship practice and practical knowledge. EE is provided only to a small number of interested people. Therefore, these areas need to be improved to increase the value of entrepreneurship. In other words, the short-term performance demands of EE due to government-led growth and the concentration of short-term evaluation systems must be improved. EE is something we should all be aware of, so it is necessary to incorporate it into the paradigm of traditional education. For successful EE, it is necessary to increase the its qualitative value based on the fundamentals of education, focusing on the characteristics of EE, trends, community cooperation, and practical content. In conclusion, although South Korea's EE has expanded quantitatively, it is essential to develop it qualitatively.

7.4.3 Defining Enterprise and Entrepreneurship

Without startups, there is no business. Therefore, in our modern society maintained by business activities, entrepreneurship is indispensable. When society takes a step forward through a period of transformation, many new types of enterprises appear. In the United States, the entrepreneurial mindset was adopted as a core task of the national innovation strategy in 2009, and the EU emphasizes the entrepreneurial mindset under the banner of "European Commission (2009) Europe has no future without an entrepreneurial mind." Global economic organizations and advanced countries, including the OECD, pay attention to entrepreneurship as a driving force for creating new values. In order to meet social demands and adapt to changes in internal and external environments as a result of the Fourth Industrial Revolution, there is a need for frequent reorganization of major subjects in higher education. Meanwhile, the government continues to propose measures to encourage youth entrepreneurship. In order to guarantee the effectiveness of the national policy and to spread the entrepreneurial spirit throughout society, it is necessary to discover and nurture outstanding entrepreneurial talents through EE.

In Korean society, starting a business has become a necessity rather than an option in the life cycle for the diversity of occupations and continuous economic activity. In order to create a university EE ecosystem, a startup-friendly university education system was implemented. And in order to create an environment for university students to establish startups, universities supported startup clubs and encouraged various startup-related efforts to improve conditions on campus and to spread awareness of startups (Ministry of Education et al., 2013). For the past several years, the South Korean government has developed various startup policies to drive job creation and economic growth, and has been conducting startup support projects such as EE, mentoring, consulting, and providing financial support (Ministry of SMEs and Startups, 2021).

Ipsos Issue Report (2019) notes that South Korea's entrepreneurship index has a relatively good "attitude," but its "ability" and future "aspirations" are relatively low. South Koreans' willingness to take risks is high, but it is analyzed that there is a lack of cultural support due to negative perceptions of companies. South Korea's risk tolerance index for startups and entrepreneurial activities is very progressive, ranking fifth in the world. However, the recognition of entrepreneurship-related opportunities fell to 41st, In addition, cultural support for startups and entrepreneurial activities was very low at 78th, so improvement is needed. In the United States, EE is provided throughout life, starting from elementary school. In particular, compared to about 25% of students receiving EE at British or American universities, the proportion of students receiving EE at South Korean universities is only eight percent (Kim, 2021). Despite the introduction of various EE programs in universities, there has yet to be satisfactory results of youth entrepreneurship. Most of the college students participating in EE are generations who did not experience EE in middle and high school. Above all, in order to revitalize youth entrepreneurship and increase the success rate of entrepreneurship, the educational environment must first be changed so that students can acquire knowledge and experiences relating to entrepreneurship.

According to a comparative study on the startup ecosystem of South Korean-Chinese college students published in June 2019 by the International Trade Research Institute of the Korea International Trade Association, the startup rate of South Korean college graduates in 2019 was only 0.8%. A main factor for this was the lack of concrete and practical preparation as well as awareness for establishing a startup. In the KISED Entrepreneurship Survey (2019), the majority of South Koreans do not consider starting a business at all. An interesting finding is that they do not try to start a business not because of their fear of failure, but that they do not even consider startup as a career. This research indicates that EE is necessary and important in South Korean society. In order to achieve technological innovation through "creative destruction," it is important to enhance and spread entrepreneurship. In particular, in order for entrepreneurship to be actively demonstrated and to achieve sustainable growth, systematic EE must be supported in the school curriculum. EE has been activated mainly in universities, while it is expanding to elementary, middle, and high schools. It is also necessary to introduce a lifelong education system of EE as a policy by expanding it to adults who have a basic education system or have graduated from college.

7.5 Internal Development of EE in HEIs

7.5.1 Development of EE Curriculum

Considering that our lives will be accelerated and changed due to the development and application of new science and technology, we need to continually update our education system. What is key for university education is to foster talents who can realize creative and innovative solutions. EE is expanding in role and scope, serving as a link to convert ideas or technologies implemented in university laboratories into products or services for our daily lives (Choi, 2017).

South Korea must move away from the traditional job application process of matching companies and talent. Through EE, innovation is needed along with expanded career education so that people can discover their passion and find their own career paths where they can show creativity and excellence. EE is the most effective means to become a technologically innovative country. The South Korean government is trying to encourage universities to improve their base for startups. Each university's career center has been reorganized to implement various policy-based support relating to youth startups, such that there is now a center dedicated to startup-related career issues.

In the study by Lee and Kim (2020), as shown in Table 7.1, the characteristics of Korean universities' career centers in different phases of development are compared and analyzed. The roles and functions of these career centers have changed since the first generation, but starting from the fifth generation, startup-related activities increased rapidly, with startup clubs launched and EE introduced. The introduction of startup work and startup courses within each university's career center is expanding. Most of the organizations related to industry-university cooperation, activation of the LINC project group, and employment and startups (Entrepreneurship Education Center) exist independently, while some of them are affiliated with the Office of Student Affairs. From 2020 onward, the university career center can be said to be in its sixth generation. Many operational changes have been made due to the Covid-19 pandemic, and the biggest is the rapid transition to digital education.

South Korea is well known for its strong IT infrastructure. Thus, the transition to virtual classes proceeded smoothly. In most cases, all classes were virtual for a while, depending on local circumstances. It went through a period of confusion because it was so different from the existing system such as the university education environment, operating system, and teaching/learning practices of students and professors, but we quickly adapted. Online education is set to continue in the future, with many changes expected in the existing university education system and operations.

7.5.2 EE Program

In Korea, there is still a lack of systematic EE programs targeting elementary, middle, and high schools. Recently, some classes with the theme of "mini company" in some middle schools were held. These were mainly conducted in specialized private high schools, and most of them depend on government-led EE programs. However, EE content is still lacking and has not been made mandatory. The Seoul Industry and Trade Promotion Agency, a subsidiary of the Seoul Metropolitan Government, in partnership with the Seoul Metropolitan Office of Education, operated a startup support program to induce entrepreneurial revitalization of high school students in Seoul and to strengthen the vocational training capabilities of frontline instructors. The Korea Entrepreneurship Foundation also develops educational content and

Table 7.1 Comparative analysis of the characteristics of employment career organizations by era and prospects for university work

Era	Comparison of characteristics of college career centers
1st generation: absent 1980s	With the advent of the employment culture, sub-cultural elements were created, and it was the period when employment career organizations were conceived. The introduction of a formal recruitment system by large corporations riding on the growth of the developing economy had a great impact
2nd generation: early stage 1990s–before IMF crisis	This is the early form of the university employment career organization. In the early to mid-1990s, as large-scale regular public recruitment began in earnest due to the external growth of large corporations, the career center appeared
3rd generation: launch After IMF crisis–2004	It was created in the process of overhauling the second-generation employment career path by universities after the IMF crisis. It was time to focus on the development of related systems and improvement of infrastructure
4th generation: growth 2004–2015	It was a time when the status of the center was upgraded as universities built an advanced model based on recommendations by the Ministry of Education. There were substantial and substantial changes, such as the reinforcement of professional manpower and the expansion of the budget
5th generation generator 2016–2019	It was an era when startups were merged into organizations as an extension of employment. As employment difficulties intensified, entrepreneurship emerged as an extension of employment, forming another axis with the Employment Pathways Organization. The startup boom acted as a direct trigger for changes in related organizations
6th generation 2020–present	The digital employment environment in the digital age caused by the Covid-19 pandemic crisis acted as a fuse that led to the launch of a new organization. The installation of programs and systems that reflect the technological elements of the Fourth Industrial Revolution occurred rapidly

Source Lee and Kim (2020)

competency tools for elementary/middle/high school EE, designs startup experience education models and manuals, and operates an online platform for youth startup experience education. The Ministry of Education financially supports an EE program for junior colleges, while the Small and Medium Business Administration's university EE provides packaged projects. Support includes startup course establishment fees, startup club development expenses, labor costs for startups, and operating expenses. Since 2012, the Ministry of Education has been promoting the strengthening of EE and commercialization of basic research through industry–university cooperation. Through the Ministry of Education's LINC project and the Small and Medium Business Administration's "Startup Leading University" fostering project, universities have each established an "Entrepreneurship Education Center" that helps develop EE, spread entrepreneurship within the university, and establish an EE system. In universities, EE can be divided into regular and non-regular courses, subject-type and non-curricular, depending on whether credits can be obtained. Regular entrepreneurship courses aim to cultivate students' entrepreneurship and management skills. Non-regular courses promote extracurricular startup activities without course credits, such as the formation of startup clubs and prototype production as well as startup-related special lectures and field tours. Student entrepreneurship supports the commercialization of ideas by developing youth entrepreneurship resources and linking the university with external networks.

The university startup infrastructure is responsible for the organization and manpower, space and equipment, and funding, such that entrepreneurship can be effectively cultivated. An analysis of the differences in startups founded by university students and those by college students shows that more than half of college students engage in startup activities while only 8.5% of university students do so. Regarding EE courses in colleges, 44.4%, 54.4%, and 1.2% of the courses are liberal arts electives, major electives, and mandatory, respectively. However, in universities, liberal arts electives and major electives account for 91.5% and 8.5%. In addition, a systematic EE and learning support program such as "ladder-type" EE content is developed, which involves learning the concept of entrepreneurship, completing a business-related major, and performing a project. For graduate students (master's and doctoral), the startup programs focus on creating new values and supporting technology and laboratory startups.

7.5.3 Cocurricular EE

7.5.3.1 EE Tools

The government, which has the greatest influence on university EE, introduced the "University Startup Operation Manual 3.0" in 2016 to raise the overall level of EE based on startup-related policies (Korea Entrepreneurship Foundation, 2019). Table 7.2 presents the consulting tools used in the manual. The methodology is a self-diagnosis model developed by the Ministry of SMEs and Startups, the Ministry of Education, and the Korea Entrepreneurship Foundation. Its "Customer-tailored University Startup Support Service" allocates scores for each of the five detailed indicators for each module.

This set of consulting tools established a system for a regular EE curriculum, joint curriculum, and the commercialization of startups that would reflect recent issues and related content. In October 2021, at the Industry–University Cooperation EXPO 2021 Korea Entrepreneurship Education Forum, the National Research Foundation of Korea published the "University Startup Management Manual 3.0+" to share the new policies for universities.

Table 7.2 EE consulting tool

Module			Summary					
			2	4	6	8	10	allotment
Regular Subject	Regular Curriculum		A course for start-ups with credits					
	Practical Start-up Training							
	Start-up transfer student ratio							
Non-Subject program	Club	Number of Club	Start-up Club, etc. Preparing for Start-up Activities					
		Student Ratio						
		Support Fund						
	Contest							
	Camp and Special Lecture							
Start-up Comercia lization	Scouting		Item Commercailization Selected Companies					
	Acceleraing		Start-up support projects, investment/specialized child care programs. etc					
	Scale-up		Growth and Global Advancement Programs(Marketing, R&D, etc.)					
	Start-up Fund Establishment and Investment		Attracting investment from selected Companies for item commercialization, etc					
	Organization		Dedicated organization, Consultation window, and integrated integration					
	Manpower		Dedicated personnel(persons), persons with more than 3 years experience(persons)					
	Space		The area of the start-up club(m2), the area of support space for start-up businesses(m2)					
	Equipment		Equipment retention, management personnel, and operation of equipment education programs					
	System	Bachelor	Start-up leave system(adoption status), Start-up specialties, Start-up scholarship, Start-up focus teachers, etc					
		Personnel						

Source Kim et al. (2020)

7.5.3.2 EE Convergence Major

The South Korean government expands university-level EE in order to develop entrepreneurial competencies to meet social needs. By launching an EE convergence major track within universities, students are provided with various learning options to increase their educational satisfaction and improve the quality of the future workforce (Im & Kwon, 2020). In particular, various academic reorganizations such as startup-related majors and convergence majors are rapidly rolled out. Im and Kwon (2020) reported that universities in South Korea aim to provide practical knowledge to students who have chosen the entrepreneurial convergence major (see Table 7.3).

Recently, colleges and universities have launched their own the EE convergence major under various names according to the strategic context of each university.

The EE convergence major track offers students opportunities to major in entrepreneurship and create practical business opportunities, in addition to theoretical education. Korea University's EE is structured into "Venture Startup Track," "Entrepreneurship Convergence Track," and "Social Venture Vitalization Track." Through its entrepreneurship convergence major, students strengthen their convergence capabilities at the academic and practical levels. Table 7.4 presents the educational goals for each EE major track and its links with participating majors.

At the same time, it is necessary to organize the major competency factors by dividing the competencies required in each track into knowledge, attitude, and skill, rather than from a job-oriented perspective. In order to revitalize the entrepreneurial convergence major, Korea University's startup convergence major allows students of various majors to jointly major in EE, which enhances the competitiveness of the university while strengthening the entrepreneurial competency of students.

7.5.4 Universities' Startup Support Organizational Structure

7.5.4.1 Startup Support Organizations

Through a dedicated organization that supervises startup activities within the university, universities cultivate entrepreneurship and provide space, facilities, equipment, lectures, and startup clubs, with the objective of reducing students' fear of failure (Table 7.5).

However, the roles of various university startup support organizations largely overlap, and thus a more efficient startup support model is needed. Considering the capabilities and resources of the university, It is necessary to establish an EE professional organization that can provide practical help in Entrepreneurship education and Startups.

As the emphasis on EE in universities increases, EE centers have been established mainly in universities selected for the LINC project. Universities that are not part of LINC may also secure their own funds to establish an EE center. The following is an example of how startup support organization is implemented.

1. Measures to revitalize the support organization for startups

    In-university startup support organization integrated model.

Table 7.3 Entrepreneurship convergence major track opening procedure

Sortation		Content	Details	Period
1	Composition of advisory committee	Track composition of advisory committee	• SME(Subject Matter Experts)/Selection of field experts	August 20–August 30, 2019
2	Environmental Analysis and Issue Deduction	Environmental Analysis	• University development plan and business analysis • Analysis of operation status	September 2–October 8, 2019
		Opinion survey of experts and education consumers(Delphi, FGI)	• Analysis of opinions of faculty, field experts, and school officials • Analysis of university student opinions	
		Environmental condition analysis Results and issues	• Derivation of implications according to environmental analysis	
3	Prior Research And advanced cases analysis	Prior research and excellent education Case analysis	• Analyze prior research for theoretical background and educational design • Analysis of domestic and overseas university startups and entrepreneurship education	October 21–October 31, 2019
4	Development Committee Review	Committee open and elicitation result	• Implications and development of trends in entrepreneurship convergence education • Designing the direction of startup education based on the current status, conditions, and vision of K University	November 1–November 15, 2019

(continued)

Table 7.3 (continued)

Sortation		Content	Details	Period
5	Major Track Development	Track Development	• Providing a track that meets social demand by combining excellent institution benchmakring, interviews with instructors, field experts, and education consumers	November 18–December 6, 2019
6	Talent Cultivation plan	Major ability	• Derivation of major competencies and learning task	November 18–December 6, 2019
		right people settled	• Establishing talent competency	

Source Im and Kwon (2020)

Table 7.4 Educational objectives

Major	Track educational objective
Venture Start-up	• Understand successful entrepreneurship patterns based on understanding the characteristics of the venture ecosystem • Cultivating professional knowledge, skills, and attitudes required for entrepreneurship execution • Cultivating the ability to discover and implement items from a convergent perspective
	<Participationg Major> Industrial Management Engineering. Business Administration. Intellectual Property. Jewelry Design. Sports Industry Management
Entrepreneurship convergence	• Learn the convergent perspective on entrepreneurship and cultivate the ability to manifest in the organization • Enhancing entrepreneurial competence to lead organizational innovation and discover new business opportunities • Cultivate entrepreneurship competency by understanding the importance of entrepreneurship in detail
	<Participationg Major> Management Information. Department of Intellectual Property. Accounting and Tax
Activating social venture	• Cultivating academic and practical knowledge level in social economy and social venture • Establish a plan to create social and economic value to activate social venture • To cultivate insight and ability to practice complex and diverse social issues
	<Participationg Major> Social welfare. youth. correctional protection

Source Im and Kwon (2020)

Table 7.5 Organizations dedicated to EE in universities

	Start-up training center	Start-up child care center	Start-up leading college (Start-up Support Group)	Entrepreneurship center
The competent authorities	The Department of Education	Small and Medium Business Administration	Small and Medium Business Administration	Small and Medium Business Administration
Year of launch year	2012	1992	2011	2014
Affiliation	LINC business group (Industrial-Academic Cooperation Group)	Industrial-Academic Cooperation Group	Industrial-Academic Cooperation Group, etc	Industrial-Academic Cooperation Group
Support target	A university student	Ordinary person	A university student, Ordinary person	Ordinary person
Major tasks	Integrated start-up education in universities Cooperation with related departments and start-ups in universities	Support for early start-ups such as space for start-ups, management skills funds, etc	Education and discovery of start-ups for university students and the general public Start-up execution support Support for growth promotion BI tenant support	Consolidation and coordination of university start-up education and support system for universities
Advantage	Specialization of start-up education Scalability by university Business links between universities	Program for supporting early stages of start-ups Obtain economic performance and international recognition	Comprehensive support from start-up education to corporate growth College students and general work	Securing manpower dedicated to supporting start-ups Support link for actual start-up education and commercialization
Weakness	Lack of support for implementing start-ups and promoting growth after training	University student exclusion Lack of links to start-up education in universities Low status in universities	Support bias for the general pubic Limits on the number of universities in operation	Lack of operating universities Difficulty in securing self-sustaining power to discontinue business support after up to six years

Source Small & Medium Business Administration and Korea Institute of Start-up & Entrepreneurship Development (2014) and Song and Kim (2020)

Integrated operation of entities related to startup support	A plan to provide one-stop support for entrepreneurship education, business incubator, commercialization, and growth support by integrating and unifying organizations related to startup support within the university into a 'Startup Support Group', etc
Applicable to	Applicable to cases where there are three or more startup support organizations such as education, business incubator, and commercialization, but there is a problem of business connection and overlapping functions for each organization
Action plan	In order to integrate startup-related organizations such as the EE center and startup incubator in the university into one organization, and to increase work efficiency and professionalism, the same person (head of the department) oversees the organization
Advantages/ Disadvantages	By unifying all startup support organizations in the university as an organization under the Startup Support Headquarters, there are the following benefits: (1) Eliminate the possibility of duplication of startup support programs and support targets; (2) Efficient operation; (3) Smooth business cooperation between existing startup support organizations. Although there are advantages such as generating synergies, if the functions of each area such as EE, business incubator, and commercialization support are not smooth, there may be confusion due to integration

The incubation center, another organization dedicated to startups, provides office space on campus for a certain period of time to prospective entrepreneurs. These individuals may have venture ideas and technologies but have difficulties commercializing their ideas due to a lack of experience and resource scarcity. Therefore, the center provides general-purpose equipment and workshops necessary for technology development. Other forms of support such as guidance and funding are also provided. The incubation center serves as a stepping stone and mentorship for early startups in order to activate them and increase their success rate. The following is a good example of revitalizing an organization dedicated to supporting university startups in Korea.

2. Linkage of organizations in charge of support for startups and regional networks

Step 1	Establishment of a 'Local Entrepreneurship Promotion Steering Committee' centered on local universities

(continued)

(continued)

Composition of the Steering Committee	The 'Regional Startup Revitalization Steering Committee' focuses on leading universities in the region, connecting the local council of the EE center, the local government EE institution, the startup support institution of the Ministry of SMEs and Startups, other support institutions, and private experts, installed inside the startup leading university
Main function	Responsible for the direction and purpose of entrepreneurship education and support for regional specialization, development of operating programs, division of roles for each institution, coordination and deliberation, etc
Operating budget	The committee operating budget comprises contributions from participating organizations
Step 2	Division of roles for each institution centered on the 'Regional Entrepreneurship Promotion Steering Committee'
Steering committee	Number of roles for deliberation and coordination of integrated operations of EE in the region

In addition, through the Startup Leading University Promotion Project, the way universities support startups has changed from multiple/distributed support to selective/focused fostering. It provides comprehensive support for programs throughout the startup process, such as nurturing pre-tech entrepreneurs and supporting EE packages. The university startup support group would form a cluster with the startup leading university as the hub. The startup leading university is chosen for its high-quality startup support infrastructure and successful startup commercialization for students and prospective entrepreneurs.

Universities that are not leading startup universities also try to strengthen the interconnection and efficiency between existing EE and support organizations through the establishment of a startup support group to comprehensively manage on-campus startup-related facilities and support institutions. Henry Etzkowitz, Chairman of International Industry-University Cooperation Association, argued that an "entrepreneurial university" gains economic independence by commercializing its research results, operating with an entrepreneurial mind.

An entrepreneurial university provides opportunities for professors and students to autonomously set their own research direction, run laboratories, and directly participate in corporate activities. In order to innovate into an entrepreneurial university, all members of the university must actively participate in change. Efforts should be made to strengthen EE in local universities, and to link the operating budget with funds and technical resources owned by the local community and local innovation agents, etc.

7.5.4.2 Startup Space Infrastructure for Students

With the goal of providing total start-up service support, some universities not only provide a startup club room and EE center, they also offer 3D printer rooms. The latter two operate as a prototype production workshop, makerspace, and workshop where students can actively pursue creative ideas using 3D printers and 3D scanners. Generally, student startup clubs provide most of the equipment free of charge. Universities with advanced startup infrastructure provide comprehensive technical support services, ranging from the prototype design stage to prototype production and reliability test evaluation.

7.5.4.3 EE Culture Support System

Several incentives are needed to encourage entrepreneurship. EE courses and activities, for instance, need to be designated as mandatory for graduation. In addition, a "startup semester" should be offered to students where they can fully focus on entrepreneurship activities. With entrepreneurship club activities, students can benefit from funding, mentoring, and special courses. In these ways, the university can create a promising startup culture.

The biggest support that student entrepreneurs can receive should be funding, including subsidies as well as costs relating to space usage and consultation. With financial support, students can avoid some of the economic risk when starting a business. Minimizing this risk makes it easier for young people to have the will to start a business. Students who start a business experience creating something out of nothing through the entrepreneurship process. With a low financial burden, regardless of the eventual success of the startup, students should take on the challenges and learn from the process. It is a good opportunity for students to acquire an entrepreneurial spirit.

In order to enhance the entrepreneurial competency of university students and to create an entrepreneurial atmosphere in society as a whole, the discovery, education, and training of entrepreneurship clubs is a key element of the university's entrepreneurial culture. When a startup club develops and discovers a novel idea, it should be able to grow its into a student startup by connecting with the financial and administrative support necessary for commercialization. In addition, successful startup representatives need to be invited to share their know-how and experience. Universities need to establish a teaching system dedicated to entrepreneurship and a talent mentoring group for startup practitioners, and provide intensive support in areas that prospective entrepreneurs mostly struggle with, such as intellectual property rights and marketing. By using the mentor pool specializing in each field by region, it should resolve the technical and management difficulties of prospective entrepreneurs and to improve the success rate of startups.

Various support programs are being conducted to spread EE and startup culture in South Korea. These include the YES Leader Entrepreneurship Special Lecture, Korea Practical Startup League, Venture Agricultural Startup Contest,

Korea Startup Competition, Korea Entrepreneurship Education Forum, World Entrepreneurship Week Event, Industry-Academic Cooperation Expo Startup Forum, University-tailored Entrepreneurship Education Consulting Program, University Faculty Empowerment Program, etc. In addition, the Ministry of Education, the Ministry of Science and ICT, the National Research Foundation of Korea, and the Korea Entrepreneurship Foundation jointly run the Student Startup Promising Team 300 Festival every year.

7.6 Approaches and Characteristics of EE

7.6.1 Expansion of Government-Led EE

South Korea's government has announced that it would make the university a base for startups and shift the paradigm of universities to becoming a startup center. The government controls and provides entrepreneurship-related support.[3] Startups are an important agenda for the government's job and economic revitalization policies. Although government support policies for youth entrepreneurship and EE have increased rapidly, in reality, youth entrepreneurship still has room for improvement. The Hyundai Economic Research Institute (2018) found that Among businesses founded by entrepreneurs in their 20s, more than half is in wholesale and retail (39.2%) and lodging and restaurant businesses (24.2%). In terms of job creation, the number of employees per young startup company in their 20s is 2.3, which is 74.2% of the average of 3.1 for all start-up companies. Financing of startups in their 20s is also very weak. The participation rate of startups in their 20s in government startup support projects is only eight percent, and most student startups do not have the competitiveness based on original technology.

With little results through merger and acquisition or initial public offering in the stock market, it is almost impossible to attract investment through accelerators, etc. Therefore, continuous interest from the central government is needed to spread EE and produce young entrepreneurs in the future. The Ministry of Education's role is particularly important as it is most involved in spreading entrepreneurship and implementing EE in HEIs in South Korea.

In order to realize the creative economy, a national goal in 2013, the Ministry of Education, the Ministry of Science and ICT and Future Planning, and the Small and Medium Business Administration announced a five-year plan for university EE. In order to encourage professors' startups and commercialization of technology, they plan to reflect the outcomes of professor startups and industry-university-research cooperation in the evaluation of teacher reappointment. As a way to revitalize university-oriented startups, the government provides on-site EE using online platforms and with local companies through university partnerships and exchanges with

[3] The Korea Forest Service holds the 'Forest Startup Idea Contest' every year.

domestic and foreign related organizations such as research institutes (e.g., Creative Economy Innovation Centers) and universities. Support for the expansion of joint business incubators and global expansion is also planned. In addition, the Korean government aims to foster technology and innovation-oriented startups in graduate schools (master's and doctoral) with EE programs for graduate students. In order to prevent the overlap of university support projects for startups, each university is encouraged to establish a startup-related "control tower."

7.6.2 Independent Development of the School of Entrepreneurship

There is a need to reorganize the university paradigm to promote the spread of technology startups. Unlike the United States, in South Korea, entrepreneurship is separated from business administration as well as the scope and results of independent research. While the various support projects of the Ministry of SMEs and Startups and the Ministry of Education focus on strengthening startup-related education programs, there is a tendency to separate EE from business administration (e.g., Entrepreneurship Education Center, Startup Support Group) within universities.

At the same time, there have been many more undergraduate and graduate EE programs. In 2004, the first Graduate School of Entrepreneurship was established in Korea under the support of the Small and Medium Business Administration. In 2007, the first doctoral program in entrepreneurship in Korea was launched at Chung-Ang University in Seoul. In 2010, Sookmyung Women's University saw its first entrepreneurship major in the undergraduate program, and in 2012, the Faculty of Entrepreneurship was established at Hoseo University (Lee & Hwang, 2015). In 2017, a Department of Entrepreneurship was established at Kunsan National University as an undergraduate course, the first in a national university. In cases where it is difficult to establish a new department, entrepreneurship convergence and entrepreneurship-related majors are offered.

For example, if you have a basic major, the entrepreneurship major would be for students who want to learn about entrepreneurship where they can systematically study six majors and four liberal arts courses. Students can study liberal arts in the field of employment and startup. In the case of extracurricular courses, practice-based programs such as EE centers, camps, and internships are offered. These are supported by the participation of fulltime professors and those in EE. In cooperation with overseas universities, Korean universities may also offer a joint course with a certificate of completion for a startup convergence major.

In South Korea, the status of EE in terms of regular university subjects, non-specialties, startup commercialization, and support infrastructure is at the stage of qualitative improvement despite quantitative performance. In order for university EE to grow qualitatively, EE modules for each level should be established, and based on this, entrepreneurship and commercialization education should be conducted step by step. It is essential to integrate EE into regular subjects such as liberal arts courses, as well as to increase the number of practical subjects and structure the contents of extracurricular subjects.

In the case of the Korea Advanced Institute of Science and Technology (KAIST), a K-School Certificate was newly established for graduate students and awarded to those who completed some K-School courses. In addition, it is necessary to actively use the systematic commercialization program of the Ministry of SMEs and Startups, and make continuous efforts to create a specialized university startup ecosystem and actual startups through cooperation with external companies.

7.6.3 Universities' Startup-Friendly Academic Support System

The startup support system in universities comprises a startup-friendly academic system, human resources system, manpower, organization, space, and equipment. Startup experts have argued that the establishment of a startup support system is the most urgent and important for university startup support activities to be internalized and advanced.

First, it is necessary to build a startup support organization, mid- to long-term roadmap, and related systems and space equipment in stages. Various startup-friendly academic systems need to be disseminated and revitalized to resolve challenges resulting from academic interruption and academic parallelism. In order to increase the effectiveness of EE and supporting activities, an entrepreneurial-friendly academic system should be established. The following is an example of a basic startup-friendly academic system in South Korea.

1. *System for taking a leave of absence to start a business.*
2. *Startup lecture unit alternative recognition system.*
3. *Startup lecture unit exchange system.*
4. *Startup scholarship system.*
5. *Startup special talent selection system*
6. *Entrepreneurship department and startup-related major.*
7. *Establishment of the operating committee for undergraduate EE system.*
8. *Entrepreneurship-friendly personnel system.*
9. *Entrepreneurship-focused professor system.*
10. *Startup research year system.*

7.7 EE Support System in South Korea

7.7.1 Stakeholders

The structure of the EE support system will be introduced in the following order.

(1) Central government universities (2) Local governments (3) Universities (4) Funding institutions (5) It is actively spreading through cooperation with stakeholders such as the private sector.

The South Korean government has been promoting entrepreneurship and EE in order to solve problems such as a sluggish economic growth and unemployment. The country's legal and institutional foundations have been reorganized to support startups at the national level, and various startup support efforts are led by different government and administrative departments. The government department that directly supports startups is the Ministry of SMEs and Startups, while the Ministry of Education provides separate support for university EE. For example, the Korea Forest Service has projects that support startup ideas such as data-based forests. The Ministry of Education, the National Research Foundation of Korea, and the Korea Entrepreneurship Foundation jointly developed the Startup Education Operation Manual 3.0+ and disseminated it to universities across the country in order to establish a standard operating model for the EE support system. According to the government's "Economic Policy Direction for 2020," startup support projects in 2020 had a budget of KRW1.45 trillion, an increase of KRW333.6 billion (29.8%) compared to 2019 (Ministry of SMEs and Startups, 2021).

In order to revitalize the economy and create jobs, the cultivation of entrepreneurship based on innovation and creativity is of paramount importance. A youth startup ecosystem should be created so that creative ideas can be derived in a virtuous cycle structure. It is necessary to nurture "ready entrepreneurs" through creative talent development and EE so that the entrepreneurial atmosphere can spread throughout society, creating jobs through the revitalization of startups. Local governments are also taking the lead in establishing a startup ecosystem for youths in order to revitalize the local economy and solve the problem of a depopulated community. In addition, the outflow of population to Seoul, the metropolitan area, and large cities is deepening. A lot of young people are leaving. Hence, local governments in Korea are making great efforts to educate the young people and support young startups in order to prevent young people from leaving. Ahn and Kang (2020) presented the difficulties faced by young people in starting a business, and provided measures and points for improvement to revitalize local youth entrepreneurship.

Another important change is the increasing participation of the private sector. In Korean society, the private sector also has a great influence on EE, establishment of startup infrastructure, and support for commercialization. With the concept of corporate social responsibility, the private sector, including large corporations, is involved in providing EE and business support.

7.7.2 EE Support Resources

7.7.2.1 EE Center

Universities have been paying great attention to EE since 2012, when the Ministry of Education launched the LINC project, which involves 61 HEIs nationwide (51 universities and 10 colleges). The establishment of EE Centers through LINC is well reflected in the policy of the Ministry of Education.

In 2012, the Ministry of Education established a government-wide cooperative system for youth entrepreneurship and implemented the Youth Entrepreneurship Support Council. Considering the overlapping issues between ministries, the task of revitalizing young entrepreneurship is divided into three stages: pre-startup incubation, startup incubation, and business growth support. In order to strengthen university EE, entrepreneurship centers have been installed in the 61 participating universities, focusing on industry–academia cooperation. Entrepreneurs are appointed as key professors of EE, mentoring students in entrepreneurship. In order to raise students' interest and cultivate an entrepreneurial mindset, universities organize roadshows and showcases of successful cases of startups. Universities also expanded support for startup clubs and EE programs through which people can form online and offline networks as well as connect with entrepreneurs and technicians. KAIST, for example, created STARTUPWON (Institute for Startup KAIST), a large-scale organization dedicated to startups.

7.7.2.2 Graduate School of Entrepreneurship

With the goal of nurturing experts in the startup field, the Small and Medium Business Startup Support Act (Article 8, Paragraphs 1 to 2) established the Graduate School of Entrepreneurship in 2004 in five regions nationwide. Another five entrepreneurship graduate schools were established in 2014. The purpose of the graduate school was to "establish an entrepreneurship education system that reflects regional specialization, nurture prospective entrepreneurs, nurture competent start-up experts, and lay the foundation for establishing a start-up infrastructure." The government provides support for instructor fees, educational development expenses, and scholarships for the graduate schools. Practical training includes domestic and overseas field training and entrepreneurship mentor training.

- *Graduate schools of entrepreneurship established in 2004.*

(1) Chung-Ang University (2) Hanbat University (3) Hoseo University (4) Yewon University of the Arts (5) Gyeongnam University of Science and Technology.

- *Graduate schools of entrepreneurship established in 2014.*

(1) Kookmin University (2) Keimyung University (3) Sungkyunkwan University (4) Yonsei University (Wonju) (5) Pusan National University.

7.7.2.3 EE University Bases

The Ministry of Education designates key universities for EE, supports other educational institutions within the region by utilizing the university's EE and support capabilities, and divides the central and southern regions to develop a leading model for EE. These form the basis for an EE support system. To establish a standard model for EE for each competency level based on the analysis of Korea's EE conditions, there are two base centers across the country carrying out three projects: (1) advancement of EE operations, (2) discovery and nurturing of startup talent, and (3) spread of university-centered startup culture.

Currently, Chung-Ang University in Seoul is the base for the metropolitan area, Chungcheong, and Gangwon. Meanwhile, Yeungnam University is the base for Gyeongsang-do, Jeolla-do, and Jeju Island. This three-year project (2020–2022) is based on the budget and promotion system of the Ministry of Education. Two universities are selected, and each base university supports an annual project costing KRW 250 million. After three years, the Center for Startup Promotion, through an open call, selects another university as a base for EE.

1. **Chung-Ang University: Center for EE in the central region**
   Chung-Ang University has been selected as a base for EE in the central region from 2020 to 2022. It has established a mid- to long-term development plan and a roadmap by establishing the Startup Common University 2025 vision for strengthening students' entrepreneurial capabilities and revitalizing EE.
2. **Yeungnam University: Center for EE in the southern region**
   Yeungnam University supports other universities and EE institutions in the three southern regions of Daegu, Gyeongsangbuk-do, Southeast, Honam and Jeju Island. It utilizes the universities' EE and support capabilities to develop a leading model for EE.

7.7.2.4 University+ Entrepreneurship Center

Major universities in Korea that operate the University+ Entrepreneurship Center instil a positive perception on the attitude toward entrepreneurship such as starting a business so that entrepreneurship can be continuously expressed in society over a long period of time. The University+ Entrepreneurship Center was established in the university as a place for education and as a cradle for planting the seeds of entrepreneurship. The University+ support system began in 2014 as a six-year project involving six universities. In 2015, three more universities were selected, and the nine universities formed a federation. Four startup-related areas have been set up within these universities: practical EE, mentoring and business incubator linkage, consulting and research, and internal/external network construction. These are implemented autonomously according to the universities' capabilities. Since the universities were selected based on autonomous business planning, various programs reflecting the university's challenging spirit are planned, reflected, and operated by the University+ Entrepreneur Center (Fig. 7.5).

Fig. 7.5 Current status of nine universities nationwide with entrepreneurship centers

The University+ Entrepreneurship Center offers a variety of regular and ad-hoc courses, on-campus and outreach programs are offered to foster and promote entrepreneurship. The target audience is wide, including not only university students but also citizens participating in the local community. The program also encourages the cooperation between organizations at universities, between universities, and with external organizations. Although it is a university-specific program among the startup-related projects of the Ministry of SMEs and Startups, it is characterized by the respective university's expertise and autonomy. The following nine universities are being supported as the Entrepreneur Centers in South Korea.

(1) Seoul National University (2) Sookmyung Women's University (3) Ewha Womans University (4) Sogang University (5) Inha University, (6) Hanyang University (7) KAIST(Korea Advanced Institute of Science and Technology) (8) Pohang University of Science and Technology (9) Yeungnam University.

Figure 7.6 shows a keyword analysis for the University Entrepreneurship Center network operations from 2014 to 2019. The six-year program focuses on network establishment through the spread of entrepreneurship.

In addition, by focusing on business model development and the strengthening of startup capabilities such as pitching strategies, the program creates a culture of early entrepreneurship. There is greater emphasis on practical startups and specialized fields. The analysis shows that efforts were geared to enhance performance. The most frequently mentioned keywords include "entrepreneurship," "cultivation," "diffusion," "network construction," "business model," "practical startup," "business model," "alumni founder networking," and "global entrepreneurship." (Fig. 7.7).

Fig. 7.6 University Entrepreneurship Center network operation keywords, 2014–2019

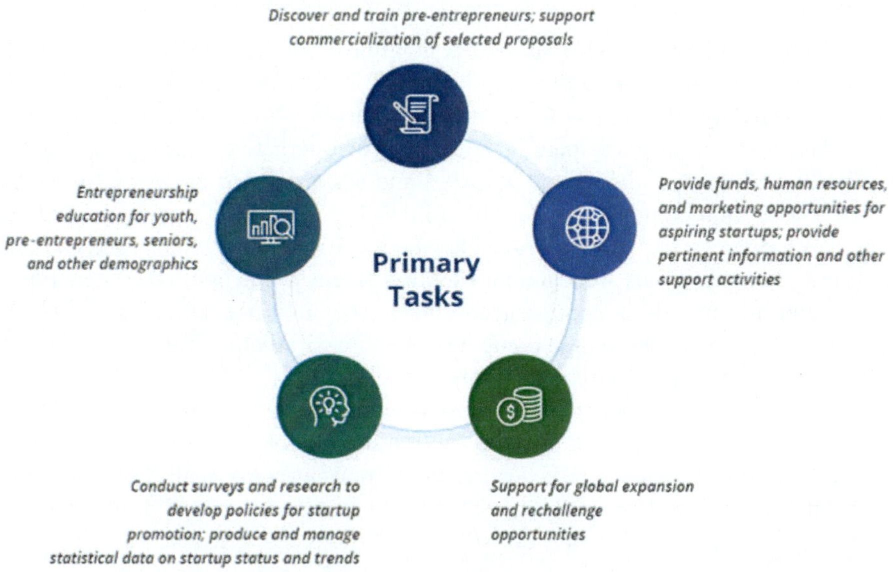

Fig. 7.7 KISED (www.kised.or.kr)

7.7.2.5 Korea Institute of Startup and Entrepreneurship Development (KISED)

KISED is a specialized agency for startup support under the Ministry of SMEs and Startups in Korea. It aims to contribute to the development of the national economy through the growth of startups and job opportunities as well as through promoting technology-based startups of future entrepreneurs by cultivating an entrepreneurial spirit.

7.8 Examples of EE Practice in South Korea

Policies of the South Korean government resulted in the rapid expansion of EE in universities in terms of quantity. Lifelong education is applied in various levels such as elementary, middle and high school, and for adults. Nevertheless, the most important group is young people. The government provides full support to revitalize youth entrepreneurship. There are various examples of EE, but the following sections will introduce cases of online EE after the Covid-19 pandemic broke out, metaverse EE, and local startups connected with local communities.

7.8.1 Case Study 1: K-MOOC and KOCW

K-MOOC[4] is a Korean-style massive open online course (MOOC) launched in 2015 that allows anyone to take any desired course for free online. K-MOOC has shifted the way learning takes place: from an online learning video where only students in the classroom could participate in, to a form of interactive learning that includes question-and-answer, discussion, quizzes, and assignment submission. In terms of digital education content, it is possible to develop systematic content according to the development technique optimized for the K-MOOC platform and the guidelines of the National Lifelong Education Promotion Agency. The optimal content type for each learning module is applied according to the subject characteristics, and it is possible to produce diversified content formats (e.g., studio, field filming, seminar, interview, etc.) to maintain the continuity of learning.

In addition, differentiated technology is used to emphasize the originality of content. K-MOOC can accommodate an unlimited number of non-specified participants. The educational course is designed to achieve learning goals. Learners' participation is maximized by applying K-MOOC-based instructional design and adult learning principles that match the subject characteristics. It also actively utilizes multiformat content including group discussions, seminars, interactions between experts and learners. There are a total of 27 startup-related courses available on K-MOOC, which is 3.6% of a total of 745 courses. Core content relating to startups is lacking. Most of the lectures are one-way, centered on explanation by the instructor, with some team teaching lectures. The course development process includes the development and provision of a manuscript writing guide that conforms to K-MOOC's requirement. Taking the purpose of K-MOOC into account, its development framework is to ensure the independence of each course, employing micro-learning production technique, create an immediate feedback environment, and smooth communication between content experts. Further, learning activities are supported by reference materials and presentation of related cases. An anticipated issue when developing digital lecture content is the risk of copyright infringement.

---

4 www.kmooc.kr

Fig. 7.8 Eight universities of the Regional University Consortium of Korea

Other public online lecture platforms where students can take online courses on startups by universities are KOCW[5] and Startup Credit Exchange. Military service is mandatory for young men in South Korea. During the service period, conscripts can take EE online lectures whose credits are recognized. Through this, the issue of disconnection in EE due to military service can be mitigated.

7.8.2 *Case Study 2: Coalition of Regional Universities Joint Online Entrepreneurship Open Lecture Production (RUCK)*

In South Korea, EE has various forms. This case is about collaborative educational content creation. In order to provide students with a variety of EE experiences, eight local private universities across the country formed the Regional University Consortium of Korea (RUCK) and have been sharing distance lectures since 2002. In this way, several universities in the region jointly develop online EE content and integrate it into the curriculum. There are cases of K-MOOC EE content. RUCK universities jointly produce EE programs and courses (Fig. 7.8).

In addition, professors specializing in each field collaborate to develop online EE courses for each degree. EE is spreading through the development of differentiated courses, including K-MOOC, and learning environments.

7.8.3 *Case Study 3: Incorporation of Metaverse Technology in EE*

In South Korea, various attempts have been made to exploit technologies of the Fourth Industrial Revolution in EE. These days, the concept of 'metaverse' has rapidly spread in various areas of South Korean society. "Metaverse" refers to an online world where people can coexist in different real spaces using virtual reality (VR) and augmented reality (AR) technologies. The term "metaverse" first appeared in Neal Stephenson's science fiction novel Snow Crash published in 1992. Recently, interest in the metaverse industry at home and abroad is increasing. This is a natural development as online culture has permeated into daily life, particularly after the

[5] KOCW: Korean Open Course Ware: Online University Open Courseware Service; www.kocw.net.

Fig. 7.9 Case of application of 'Metaverse Gather Town' system to EE

spread of Covid-19. Many games are based on a specific worldview, and economic and social activities are built with digital data, and online culture is a part of daily life for South Korea's MZ generation.[6] The importance of the metaverse is increasing as the MZ generation will become the backbone of society in the future, and the proportion of online social culture will intensify in the future due to the pandemic. Recently, South Korea's Soon Chun Hyang University has received attention for introducing the metaverse tool at its welcome party and entrance ceremony for new students.

As college life is based on dialogue and communication, novel attempts are being made to apply the metaverse to college education and life. South Korean EE experts and professors are quickly learning about the metaverse system. EE classes at universities will be transformed to combine virtual and real worlds. In EE, communication and collaboration are very important. But after the coronavirus pandemic, it was almost impossible to meet and work together. Entrepreneurship revitalization naturally stagnated, and there were many difficulties in conducting EE. However, now, by incorporating technology such as metaverse, changes are being made to EE. Sharing and collaboration are keywords for the future direction of university education (Fig. 7.9).

To this end, various communication methods are required beyond online education, and the metaverse technology is being applied to EE, where networking is key. Metaverse is used in Korea's EE field as a tool to overcome spatial limitations and communicate with each other beyond regional boundaries. As the most representative example, Metaverse was used in the 2021 Ultra-wide Metaverse Idea Stone startup contest of the LINC+ program, which consisted of seven universities. The metaverse system supports the commercialization of students' new startup ideas,

[6] Millennials born in the early 1980s to early 2000s and Generation Z born in the mid-1990s to early 2000s. They are familiar with the digital environment and show the characteristics of pursuing the latest trends and a unique experience different from others.

discovery of startup innovation growth platforms, revitalization of EE, and strengthening of industry–university cooperation through mutual exchange between universities across different regions. In the era of globalization, metaverse is expected to bring various opportunities and changes to university EE.

7.8.4 Case Study 4: Cases of EE and Startup Support for Large Companies in the Private Sector

Private sector conglomerates are supporting EE, startups, and startup infrastructure in various fields in South Korea. Samsung, for instance, has consistently supported in-house EE programs by launching its in-house venture fostering program known as C-Lab Inside in 2012. It also hosts international nonprofit youth educational institutions such as JA Korea and Startup Playground Festival, which supports youth entrepreneurship.

Other large corporations such as SK and Hyundai have also been actively supporting youth startups. SK has long supported the social economy sector and nurtured social ventures and social enterprises. In particular, SK E&S in the energy sector is attempting to solve problems in the local community with Localize Gunsan, an urban regeneration project, and is revitalizing local startups through its corporate social responsibility activities. As startups that have grown through urban regeneration projects succeed in commercializing their products and services, their entry into the market is gradually expanding. Local: Rise Gunsan is a project to revitalize Gunsan, Jeollabuk-do, which was once an industrial city centered on automobiles and shipbuilding, but has recently contracted. The main goal is to develop the old downtown Yeonghwa-dong area into a cultural and tourism center of Gunsan and to revitalize the city by creating local jobs. Innovators with novel ideas are selected and provided with EE, business startup practical education and support, activity expenses, product development, and public relations. As such, large corporations are also fulfilling their social responsibilities by fostering start-ups. In the case of Hyundai, through a program called H-OnDream, various social enterprises are nurtured and new values are created through EE and startup support.

7.9 Challenges and Future Directions for EE in the Republic of Korea

Fundamentally, EE is a part of entrepreneurship practice, distinguished from entrepreneurship. Even in universities, EE is generally misunderstood as practical education for business startup practice. It is necessary to distinguish between basic education, which provides the necessary competencies to navigate one's own life, and practical education for entrepreneurship, which creates jobs with creative ideas. Both basic and practice-based education are important educational systems that a university should provide. Beyond the traditional purpose of education and research, universities these days are responsible for economic growth (Hoskisson et al., 2011). An important role of the university is to nurture talent and conduct research. Entrepreneurship activities provide opportunities for students to acquire the entrepreneurial spirit. Further, as shown in previous studies, students who have a high level of entrepreneurial spirit are more likely to grow and succeed as human capital that contributes to the society. Since students with a high probability of success in society can be competitive advantage, entrepreneurship activities can help universities to nurture these personal traits. The tasks and future directions of entrepreneurship in Korea are summarized as follows.

First, it is necessary to improve the quality of EE. In the past few years, the entrepreneurship environment of Korean universities has improved remarkably, but there is still room for improvement in EE. The challenge of starting a business at a university has not been transformed into the mainstream culture where there is actual job creation. Although EE has increased in quantity, most students think that they are separated from entrepreneurship, and a significant number of students who experience EE use their entrepreneurial skills attained for employment rather than for their own entrepreneurial activities. They think of it as additional activities considered valuable for employment. Along with the quantitative growth of university startups, qualitative growth of university EE and related support systems is necessary so that entrepreneurs who challenge innovative youth startups with high added value can be produced (Kim, 2017). The specialization of entrepreneurship, improvement of the quality of EE, reorganization of startup subjects, and establishment of a systematic EE roadmap for each stage of education are necessary from a policy point of view. In conclusion, in order to improve the quality of EE, various educational methodologies are needed such as textbook development, games, metaverse, and non-face-to-face systems, strengthening practical education.

Second, EE needs to be converted into basic mandatory education. EE should be conducted with the concept of lifelong education from the perspective of life cycle, rather than education focused and limited to a specific group (e.g., nascent entrepreneurs). Universities should spread entrepreneurship in the fields of education, research, and industry–university cooperation to all departments and members (professors, staff, and students) within the university, and expand its links with the local community.

EE should be approached from the perspective of lifelong education. It is important to design EE for each stage of education that fits the characteristics of universities and junior colleges, rather than designing EE that depends on the Ministry of Education's comprehensive university evaluation. With the concept of lifelong learning, it is necessary to develop a curriculum and teaching materials for EE from elementary school to adult learning through EE programs for each educational stage.

Third, it is important to have experts in EE, practice, and support within the university. Improvement of the competency of experts who provide startup-related education or programs is necessary to ensure the quality. While interest in EE has been increasing in South Korean society in recent years, the training of educators is insufficient. It is difficult for startup experts to properly settle in the university because many casual employees have their own business, the teachers do not hold professions in entrepreneurship field, or they do not match the requirements and regulations of the existing professor recruitment standards. Universities in South Korea are conservative in terms of change and innovation. In order to recruit talented startup experts to universities in the future, university recruitment regulations and teaching culture should be more open. Educational requirements of students have diversified and the educational environment has changed, but the university operating system cannot keep up with the trends of the times. In addition, a long-term systematic plan should be prepared for nurturing professional manpower in EE.

As seen in examples from other countries, most of the professors who teach EE at prestigious universities have practical experience in entrepreneurship and there are researchers who have studied entrepreneurship for a long time. If the proportion of outsourced professors or instructors increases in the short term due to the importance of practical work, it may be hard to establish systematic management practices and low teaching quality is expected. Therefore, it is necessary to manage the advanced management of the quality of EE by hiring a startup experts centered on practice and research as a fulltime professors.

Fourth, to develop and produce results, university EE needs cooperation from stakeholders (e.g., friends, parents, professors, etc.) who have the most direct influence on the career path of university students. For example, we will need to educate those stakeholders and make them aware of the importance of EE. According to a 2017 survey by the Presidential Committee on Youth, 28.1% of 423 young entrepreneurs stated that their parents were against starting a business, followed by "because I wanted to find a stable job" (37.8%) and "because I thought it was difficult to succeed in starting a business" (22.7%). Ahn et al. (2017) reported that social support through parental cooperation is essential in order to increase the will of university students to start a business.

Fifth, universities should be transformed into an entrepreneurial spaces where various innovative startups can take place on campus. It should create a culture that encourages entrepreneurship on campus. The South Korean government plans to designate and support universities with high-quality startup support systems, infrastructure, and youth environment as "startup-oriented universities" in the second half of 2021 in order to strengthen the role of universities. Therefore, it is necessary to share the resources (human, physical, and spatial) of the university by opening up the campus to the community. Students can realize the theories learned in entrepreneurship courses to come up with creative ideas, use the university's equipment to produce prototypes and finished products, and apply technology, products, services, etc. to the campus. To build an entrepreneurial university, innovation is needed. At the national level, the role of universities is imperative for creating new jobs and new growth engines, as well as creating venture startups with creative and innovative ideas (Kim, 2013).

Sixth, global competency should be nurtured through EE. A global mindset should be cultivated, and a support system established with overseas university students to exchange startup ideas and cultivate a global mindset through global EE. In the future, it is necessary to collaborate with major global universities in EE, create a global startup ecosystem, and foster global startups. In 2021, President Kwang Hyeong Lee of KAIST declared its "1LAB 1STARTUP" plan, that is, one lab would produce one startup company. For reference, there are about 700 Research Lab at KAIST. The university is making efforts to ensure that the technology it possesses is not limited to research, but can also lead to commercialization and Deep Tech & Deep science startups.

Also, nurturing successful entrepreneurs is not an easy task: entrepreneurial spirit and entrepreneurial ability are not formed in the short term. However, Korea's EE is currently being evaluated based on short-term results from government-led startup support. If it continues to lead to short-term effects of government financial support, it will be difficult to establish a proper startup ecosystem and startup culture. In order to overcome these problems, a mid- to long-term approach is important.

Finally, I would like to make a new proposal to the Entrepreneurship Education System in order to spread Entrepreneurship and activate Startups. It is also necessary to induce the active participation of stakeholders and appropriately reward contributors for nurturing successful entrepreneurs. Therefore, I would like to propose the **"Successful Entrepreneur Development Solidarity Contribution System."** In other words, nurturing successful entrepreneurs has the effect of creating various added values such as revitalizing the national economy, creating jobs, creating value, and increasing taxes.

It is a system that rewards achievements in nurturing actual entrepreneurs through a system in educational institutions. The government provides financial support for the operations of educational institutions and schools (universities, high schools, middle schools, and elementary schools) that have nurtured and produced entrepreneurs who have reached the stage-by-stage startup success evaluation index. It is proposed to create a system that values contributions to fostering successful entrepreneurs, which then incentivizes universities, high schools, middle schools, and elementary schools. Then, it will be possible to induce voluntary and active participation and interest in university advisers and faculty evaluation scores as well. In the future, EE and entrepreneurship will be an essential part of our society. Efforts to improve the quality of EE are needed. The key to reviving university education, which has fallen into the "trap of standardization" for future generations, is to establish a system that fosters creative talent through EE along with major education and innovate into an entrepreneurial university.

References

Ahn, T. U., & Kang, T. W. (2020). A study on difficulty factors of youth startups for activating local startups. *Asia-Pacific Journal of Business Venturing and Entrepreneurship, 15*(2), 67–80 [In Korean].

Ahn, T. U., Lee, I. H., & Park, J. W. (2017). A study on the effects of creativity competency education on self-efficacy and entrepreneurial intention: The moderating role of social support through parent cooperation. *Asia-Pacific Journal of Business Venturing and Entrepreneurship, 12*(6), 25–39 [In Korean].

Ahn, T. U., & Park, J. W. (2018). The effect of entrepreneurship education on the career path of university students. *Asia-Pacific Journal of Business Venturing and Entrepreneurship, 13*(2), 177–192 [In Korean].

Bailetti, T. (2011). Fostering student entrepreneurship and university spinoff companies. *Technology Innovation Management Review, 1*(1), 7–12.

Ban, S. S., Lee, H. S., Choi, S. L., & Jeong, D. S. (2008). Korean entrepreneurship development training program: Focusing on entrepreneurship MBA courses. *The Korea Entrepreneurship Society, 3*(2), 1–42 [In Korean].

Choi, J. I., & HwangBo, Y. (2012). Development on educator for entrepreneurship program. *Asia-Pacific Journal of Business Venturing and Entrepreneurship, 7*(1), 207–214 [In Korean].

Choi, J. I., & Yang, Y. S. (2012). Program development for entrepreneurship education. *Asia-Pacific Journal of Business Venturing and Entrepreneurship, 7*(1), 125–134 [In Korean].

Choi, S. B. (2017). Career education and entrepreneurship education in universities in the era of the 4th industrial revolution. *Educational Development, 44*(3), 36–43 [In Korean].

European Commission. (2009). *2009 entrepreneurship in the EU and beyond*. Flash Eurobarometer N° 283 (Analytical Report).

Global Entrepreneurship Monitor. (2020). Retrieved from https://www.gemconsortium.org/economy-profiles/south-korea-2/policy

Ha, G. S. (2014). How do I get to college start-up support? What is the problem with college start-up support programs? *Higher Education, 185*, 18–24 [In Korean].

Han, K. S. (2007). The systematization plan of entrepreneurship education. *Korean Business Education Review, 47*, 379–405 [In Korean].

Heo, S. Y., Jang, H. E., & Lee, J. H. (2017). Exploring the government policy for promoting university startups in Korea. *Journal of the Korea Academia-Industrial Cooperation Society, 18*(1), 75–84 [In Korean].

Hong, J. D., & Gang, H. G. (2016). A study on the current status of startup education in Korean university: Focused on undergraduate and graduate degrees and non-degree courses. *The Korean Society of Business Venturing*, 163–165 [In Korean].

Hoskisson, R. E., Covin, J., Volberda, H. W., & Johnson, R. A. (2011). Revitalizing entrepreneurship: The search for new research opportunities. *Journal of Management Studies, 48*(6), 1141–1168.

Hyundai Economic Research Institute. (2018). Tasks and implications of social entrepreneurs in their 20s. *VIP Report, 688*, 17–15 [In Korean].

Im, J. H., & Kwon. H. (2020). A study on the establishment of an integrated model for the concept of entrepreneurship. *The Journal of Humanities and Social Science 21, 11*(6), 1395–1410 [In Korean].

Ipsos Issue Report. (2019). *Current status and implications of 'entrepreneurship'* (Korean society. No. 47).

Jung, C. G. (2013). Current status and tasks of university start-up education for the revitalization of young entrepreneurs. *The HRD Review, 16*(5), 40–59 [In Korean].

Kim, C. H., & Yang, D. W. (2018). A study on effect of the university student's entrepreneurial education participation on entrepreneurial intention: Focusing on mediation effect of entrepreneurial education satisfaction and moderate effect of startup support system. *The Korea Entrepreneurship Society Conferences, 1*, 50–75 [In Korean].

Kim, J. H., Lee, S. H., & Nam, J. M. (2020). A case study on the advancement of uni. start-up education. *Journal of Convergence for Information Technology, 10*(6), 96–108 [In Korean].

Kim, J. S. (2017). Qualitative growth of university start-ups is necessary. *KDI* , Column No.11, 86 [In Korean].

Kim, K. H. (2021). Entrepreneurship: A solid entrepreneur is the driving force of the national economy. *KDI*. SPECIAL THEME Column issue 4. *365*, 18–19 [In Korean].

Kim, S. W. (2013). A method for establishing a university's entrepreneurship-friendly system to promote student and faculty entrepreneurship. *Science and Technology Policy, 191*, 34–41 [In Korean].

KOEF (Korea Entrepreneurship Foundation) Survey. (2018). *2018 entrepreneurship survey'* (Result of a survey of 12,038 general citizens aged 13 to 69) [In Korean].

KOEF (Korea Entrepreneurship Foundation). (2019). *University startup management manual 3.0+. Report* [In Korean].

Korea Institute of Startup & Entrepreneurship Development. (2019). *2018 University startup statistics survey results report.* Retrieved from https://www.kised.or.kr [In Korean].

KOREA Venture Business Association. (2020). *2020 Venture business precision survey* [In Korean].

Korean Economy Business. (2019, March 26). Part 1 In search of forgotten 'entrepreneurship'—Hankyung Business 600 people Survey... ."Entrepreneurship education required" 87.3%

Lee, J. G., & Kim, H. W. (2020). A study on the comparative analysis of the transition process and characteristics of employment & career organization in Korean universities by the period. *The Review of Business History, 35*(3), 45–63 [In Korean].

Lee, W. J., & Hwang, B. Y. (2015). The chronology of Korean entrepreneurship education: The feature analysis and future development of entrepreneurship education. *Asia-Pacific Journal of Business Venturing and Entrepreneurship, 10*(3), 171–183 [In Korean].

Lee, W. J., Kim, J. S., & Kim, Y. T. (2013). An empirical study on the influence of entrepreneurship education quality and satisfaction to university student's entrepreneurial intention. *Korea Association of Business Education, 28*(3), 105–131.

Lee, Y. D. (2013). University-based tech-startups. *IE Magazine, 20*(2), 14–22 [In Korean].

Lee, Y. H., & Park, S. H. (2014). A study on the success factors of venture entrepreneurs and entrepreneurship education. *Asia-Pacific Journal of Business Venturing and Entrepreneurship, 9*(6), 231–244 [In Korean].

Lee, Y. J. (2013). How to revitalize start-ups in the age of creative economy. *Science & Technology Policy, 191*, 10–21 [In Korean].

Lerner, J. (2000). The government as venture capitalist: The long-run impact of the SBIR program. *The Journal of Private Equity, 3*(2), 55–78.

Ministry of Education. (2018). *University-industry-university cooperation research report* [In Korean].

Ministry of Education, Ministry of Science, ICT and Future Planning and Ministry of SMEs and Startups. (2013). *Five-year plan for start-up education in universities press release.* [In Korean].

Ministry of SMEs and Startups. (2021). Press release. *Crisis is an opportunity for entrepreneurs, Korea rises to 9th in the world in the Entrepreneurship Index'* [In Korean].

Mok, Y. D., & Choi, M. G. (2010). A study on curriculum design of entrepreneurship education in undergraduate school. *Journal of the Korea Academia-Industrial Cooperation Society, 11*(1), 320–334 [In Korean].

Ryu, J. S., Kim, J. H., Park, S. I., Jo, J. S., & Jung, H. J. (2014). *A study on the employment impact assessment of youth start-up through the five-year plan of university start-up education.* Ministry of Employment and Labor & Korea Labor Institute. Report [In Korean].

Song, Y. M., & Kim, Y. S. (2020). Analysis of entrepreneurship education in Korean and American universities. *The Korean Society Fishries and Sciences Education, 32*(4), 988–1006 [In Korean].

Statistics Korea. (2020). *Trends in the youth employment rate.* Retrieved from http://www.index.go.kr [In Korean].

Yoon, E. J. (2018, May 25). Strengthen university entrepreneurship education: Target of 5,000 student start-ups. *Veritas Alpha.* [In Korean].

Yu, In Young (2022). The crisis of local universities is the crisis of local extinction. *Happy Education*, Vol. 4 [In Korean].

Chapter 8
Entrepreneurship Education in the United Kingdom

Robin Bell

8.1 Introduction

Entrepreneurship education (EE) has been present in some form in UK higher education since the early 1970s. Over the last 50 years, both its scope and prominence have increased with more emphasis on EE as a distinctive discipline. This is grounded in the belief and understanding that entrepreneurship, or at least certain facets of it, can be taught (Kuratko, 2005). Therefore, individuals who acquire both the right knowledge and skills are in an improved position to start, develop, and grow a successful business. Against this backdrop, successive governments have promoted entrepreneurship and EE within higher education. The UK Quality Assurance Agency for Higher Education (QAA), which supports, monitors, and checks quality standards within the UK, has issued guidelines about the aims and delivery of effective EE and sought to define some terms. However, the delivery and provision of EE largely lies at the institutional level, where individual institutions have the autonomy to tailor courses to suit their student cohorts. Institutions and educators adopt a range of methods and approaches for the delivery of EE, underpinned by different educational philosophies and theories to inform their pedagogic practice. Whilst the QAA (2018) guidelines make a distinction between entrepreneurship and EE, this chapter uses the term "EE" more generally in line with other parts of the world, so as not to confuse readers who may not be familiar with specific UK definitions. The distinction between enterprise and EE also remains somewhat contentious within existing literature (Bridge, 2017), which will be discussed later in the chapter.

The chapter will first explore the historical development of EE in the UK, before discussing the current guidelines. Following this, a range of current approaches to EE and the underpinning educational theory and philosophy are discussed. The penultimate section provides two cases of the delivery of EE in the UK, which are linked

R. Bell (✉)
Worcester Business School, University of Worcester, UK
e-mail: r.bell@worc.ac.uk

X. Xu, *Comparative Entrepreneurship Education*,
https://doi.org/10.1007/978-981-99-1835-5_8

to the previous sections on guidelines and the underpinning educational theory and philosophy. Finally, the chapter concludes by discussing the challenges and future directions for EE in the UK.

8.2 Historical Development of EE in the UK

The roots of EE in the UK can be traced back to business schools in the early 1970s. Traditionally, business schools within the UK were focused on management education and developing professional management competencies and skills. From the early 1970s, there was a gradual increase in entrepreneurship modules within degree programs, which were focused on entrepreneurship and small businesses. These modules were often isolated within curricula, and at this time in the UK there was limited distinction between EE and small-business education. In contrast, in the US, EE was targeted at high-growth new ventures and taught at more prestigious institutions whilst small-business education was targeted at running small businesses with only limited growth potential and was taught more widely (Watkins & Stone, 1999). Within the UK, the early development of small-business education and EE grew within both prestigious and vocational institutions, with representation at both "red brick" universities and polytechnic institutions. However, supporting enterprise and innovation was more challenging in the vocational and polytechnic settings due to more limited resources, which led to development in this area being stifled. EE and training were largely reserved for postgraduate and owner-manager audiences, rather than being open to all students (Watkins & Stone, 1999). This was based on the belief that undergraduates would have limited benefit from, and interest in, small-business and entrepreneurship education. Additionally, there were challenges in scaling up what at the time were specialist courses to wider audiences.

A major shift in the views on entrepreneurship and EE emerged in the second half of the 1970s. The Sterling Crisis of 1976, which pushed the value of the pound down and forced the UK government to borrow from the International Monetary Fund, led to an economic downturn and a growth in unemployment (Newman & Giles, 2005). To try and mitigate the rise in unemployment, the UK government invested in delivering entrepreneurship training and courses, which were called New Enterprise Programs, for managers who had worked in large organizations and found themselves unemployed. Whilst such initiatives are now commonplace, this represented the first time UK public funds were used in such a way (Watkins & Stone, 1999). These courses continued to develop, evolve, and grow through the 1980s and early 1990s to meet the changing economic needs. This led to some differentiation and funneling between courses and programs, depending on the growth potential of the entrepreneurial idea. Potential high-growth businesses were steered toward major business schools and less innovative potential ideas were steered toward polytechnics. Meanwhile, older and more prestigious universities often stayed out of entrepreneurial training and support, and instead only engaged from a research standpoint (Watkins & Stone, 1999). Eventually, in the 1990s, these courses, and programs were taken over by local enterprise partnerships and councils, allowing them to become more localized. During this time, higher education had also strengthened their EE and entrepreneurship support provision, with faculty, research, experience, and expertise growing in these areas.

Toward the end of the 1980s and 1990s, the UK saw a fast expansion and growth in higher education. This period also saw the dissolution of the binary divide between polytechnics and universities (Mayhew et al., 2004). As more students were entering UK higher education, there was an increased emphasis on the outcomes of higher education, including employability. To ensure value for money and future funding, institutions needed to demonstrate that they offered value for money in terms of developing skills and employability. Cranmer (2006) argued that ensuring the employability of graduates was central to contemporary UK higher education and that this focus on employability led to new teaching approaches and strategies as well as the introduction of new courses explicitly delivering employability skills. As students and graduates with enterprising skills and competencies were generally considered to be more employable (Rae, 2005), there was increased emphasis on developing these abilities through EE. The role of universities to support entrepreneurship was further crystallized by the National Committee of Inquiry into Higher Education's (1997) recommendation that universities consider the scope for encouraging entrepreneurship through innovative approaches to course and program design. Building on this, in 2000, Universities UK made business and entrepreneurial development one of the four strategic goals for universities. The emphasis on enterprise and EE continues to grow with the UK Higher Education Funding Council emphasizing that entrepreneurship should be incorporated and embedded throughout all disciplines of higher education (Higher Education Funding Council Executive, 2004). Such policies and provisions were introduced as it was perceived that higher education institutions (HEIs) were critical for the success of generating and developing entrepreneurial ideas and business talent (Robertson & Collins, 2003). At the same time, the requirement for EE was linked to and underpinned by the government's emphasis on graduate enterprise and its value and importance to the national economy (McKeown et al., 2006). Through higher education, it was expected that students and soon-to-be graduates would have access to the requisite knowledge-based resources for supporting sophisticated enterprises that could compete internationally (McMullan & Long, 1987). Several initiatives were implemented to support entrepreneurship and graduate entrepreneurship, including the establishment of 13 UK Science Enterprise Centres and consortia in 2001, funded by the UK Department for Trade and Industry. This consortium was later revised when the original funding concluded, with membership becoming open to all universities and the organization rebranded as Enterprise Educators UK (EEUK) in 2007 to reflect its wider remit and reach (EEUK, 2021). EEUK membership was opened beyond higher education to also include further education colleges and other organizations with a clear EE purpose in 2008. Membership continued to grow, reaching 100 members in 2014 (EEUK, 2021). In 2004, the government established the National Council for Graduate Entrepreneurship (NCGE), with the aim of raising the profile of entrepreneurship and promoting startups as a career choice amongst students and graduates. To achieve this, the NCGE sought to develop and promote a culture of entrepreneurship within higher education through research, education, and facilitation (Pittaway & Cope, 2007). NCGE later became known as the National Centre for Entrepreneurship Education (NCEE). The government support and intervention provided during this period to promote and support entrepreneurship and EE highlighted and reaffirmed the perceived value and role that higher education can play in encouraging graduates to start a business (Hannon, 2006).

Government calls continued for universities to develop as entrepreneurial institutions that could support the development of entrepreneurial ideas and exploit the creative potential, skills, and knowledge within higher education. Through the delivery of EE, it was expected that both students and faculty could be encouraged and would be more likely to set up new ventures and support continued growth in new businesses (UK DfES, 2003). The direction and focus on enterprise and EE led to UK universities moving toward the concept of "entrepreneurial university," where universities need to remain flexible to strategically adapt to opportunities to support and play a role in regional development (Gibb & Hannon, 2006). This challenge for universities to adapt was not only a UK challenge, but also one that universities in North America faced (Doutriaux & Barker, 1996; Schramm, 2006). This emerging role of universities has continued to be part of the dichotomous functions of modern universities in the UK, where universities are expected to be entrepreneurial and focus on both entrepreneurship and innovation to contribute to economic growth and competitiveness (Audretsch, 2014; Urbano & Guerrero, 2013). NCGE (2004) concluded that bringing entrepreneurship and EE into higher education led to a vibrant cultural shift and had a profound impact on HEIs.

Matlay and Carey (2007) conducted a 10-year longitudinal analysis of EE initiatives in the UK from 1995 to 2004. The study found that all 40 universities in the sample provided at least some small-business and EE courses. During the period of 1995–1999, EE provision was greater at new universities, but the gap between EE provision at new and old universities was reduced during the second five-year period of 2000–2004. In the latter period, the overall amount of EE provision grew considerably. By the end of the 2004 academic year, most of the sample universities were offering EE at both the undergraduate and postgraduate levels. The audience to which EE was delivered also extended substantially during the period of the research, with most of the offerings provided to business school students in the 1995–2000 period, and the provision of EE extended outside of the business school in over half of the sample during 2000–2004. Where EE was not offered to all students both fulltime and part-time and undergraduate and postgraduate, a lack of demand from students was cited as the most common reason, followed by a lack of interest amongst staff, and a shortage of funding. As EE is not mandatory in UK higher education, it is driven by student demand and the capacity of HEIs to staff and resource the delivery of such modules. In Matlay and Carey's (2007) research on the nature of EE delivered, they found that there were no significant commonalities in the conceptual approach to EE. In effect, each university decided upon and used its own definition as to what constituted EE. This meant that there was "pragmatic fluidity" in terms of both the concept of EE and how it was delivered. It was determined that EE had commonly transitioned from being dominated by a variety of traditional courses borrowed or adapted from general business education to more specific, tailored EE. In the majority of the cases in the sample, the assessment in EE courses still followed traditional assessment patterns, i.e., subject coursework and end-of-term exams.

Despite the development and integration of new EE and support programs into UK higher education, some questions remained and became more pertinent, such as the operational definitions of enterprise and entrepreneurship, the underpinning pedagogy to effectively teach EE, and how EE learning should be assessed (Pittaway & Cope, 2007). The next section will explore the current guidelines and policies relating to EE in the UK.

8.3 Current EE Guidance and Policies

In common with most developed countries, the fine details of enterprise and EE policy in the UK tend to change with each new government. However, the emphasis on EE in general has remained relatively consistent, possibly demonstrating the recognition that EE has social and economic benefits (Henry, 2013). Policymakers in the UK and Europe have posited that EE can produce skilled graduates who can positively influence society and economic growth by creating jobs and value for society, through applying creativity, problem solving, innovating, and identifying business opportunities (Department for Business Enterprise & Regulatory Reform, 2008; European Commission Enterprise and Industry Directorate-General, 2008; QAA, 2018). However, critical questions have emerged within the literature as to whether policymakers are expecting too much for EE and whether it has been wrongly perceived as a panacea for solving wider societal and economic challenges (Henry, 2013).

Entrepreneurship has been supported through numerous streams including the general promotion of entrepreneurship, educational and training initiatives, and making startup funding available for entrepreneurs. The general promotion of entrepreneurship has been conducted through initiatives such as enterprise week and annual global entrepreneurship week, which acquaint people with the concept of entrepreneurship and positively adjust their perceptions of entrepreneurship. Enterprise and entrepreneurship have been promoted in both general education and subject-specific curricula such as science, engineering, and technology to highlight the value of entrepreneurship and entrepreneurial opportunities. Advance HE and its precursor, the Higher Education Academy, which promotes and advocates quality evidence-based teaching methods and developed the UK Professional Standards Framework for higher education practitioners, has supported the introduction and development of entrepreneurship into higher education curricular. Jones (2014) suggests that entrepreneurship and enterprise had gradually found their way into the UK higher education curricula, and this was achieved with an increasing focus on embedding entrepreneurship within nonbusiness disciplines including vocational and professional programs. This was based on the premise that every student, regardless of their intended career path, should have the opportunity and be encouraged to engage with entrepreneurship, which could help create and foster an enterprise culture in new graduates (Henry, 2013). As a result, entrepreneurship is increasingly being viewed within higher education as relevant to everybody rather than a select few with the best ideas (Bell & Bell, 2016a; O'Brien et al., 2019).

Whilst there have been, and currently are, numerous initiatives to support and encourage EE in higher and further education, it is not a formally mandated part of the curriculum. However, guidance on EE in the UK is provided by the Quality Assurance Agency for Higher Education (QAA)—an independent body that checks quality standards, conducts quality reviews, and develops and provides reference points and guidance for institutions. The QAA's (2018) *Enterprise and Entrepreneurship Education: Guidance for UK Higher Education Providers* provides guidelines and a benchmark for educators and institutions around which to build their EE provision. This document was an enhancement of the guidelines provided in the previous iteration in 2012. The QAA (2018) guidelines address some of the questions previously raised and issues identified around the value and role of EE, the operational definitions of enterprise and entrepreneurship, how EE can be effectively delivered, and how learning should be assessed (Pittaway & Cope, 2007). The guidance on these areas will now be discussed and summarized.

8.3.1 The Value and Role of EE

The guidance provided by the QAA (2018) posits that enterprise and EE offer the opportunity to support the development of behaviors, attributes, and competencies that are likely to have a positive impact on students' careers. This in turn can have a positive effect in terms of economic, social, and cultural value. It recognizes that enterprise and EE can be an effective tool to prepare students for the changing and potentially challenging environments that they will facc in their careers. This view of enterprise and EE is not solely focused on employment, but acknowledges that the skills and competencies gained from EE can support students and graduates to live rewarding and self-determined professional lives.

The view of enterprise and EE adopted by the QAA (2018) espouses the value of enterprise and EE for all students, not only those currently with entrepreneurial ideas and interest in setting up a business (Bell & Bell, 2016a; O'Brien et al., 2019). Instead, developing entrepreneurial competencies in graduates can offer value to the organizations they work for in the future by supporting them to be innovative and to remain competitive, thus bringing economic and social benefits. This view also recognizes that enterprise and EE can support not only graduates in developing new ventures, but also support them in their career prospects. This is underpinned by previous research that has found that entrepreneurial skills and competencies support an increased chance of employment in a professional or managerial job after graduation (Bell, 2016a, 2016b). The UK government has encouraged HEIs to develop entrepreneurial, innovative, and highly employable graduates. This has led to enterprise and EE being placed high on the agenda of HEIs (Sewell & Pool, 2010).

The QAA (2018) also highlights the wider benefits that engaging with enterprise and EE can bring to HEIs. Beyond teaching and learning, activities such as research and knowledge exchange can help universities engage with industry and communities. Therefore, enterprise and EE are an important element of the higher education landscape for developing entrepreneurial expertise and closer links with industry. This views HEIs as a potential catalyst for regional development by educating people and attracting well-educated people, facilitating knowledge transfer, and contributing to the development of new ventures and maintaining the competitiveness of existing businesses (Klofsten et al., 2019).

Tensions within the conceptualization of EE and what it should aim to achieve and cover have arisen from a lack of definition, so the following section explores the definitions presented in the QAA (2018) guidelines.

8.3.2 Defining Enterprise and Entrepreneurship

The term "entrepreneurship" is commonly used in a generic and interchangeable fashion, in a variety of contexts and for explaining many things (Matlay & Carey, 2007). The diversity in the definition has been identified within existing literature, but the search for conceptual and contextual convergence has only resulted in increasingly complex discussions across numerous disciplines that offer different definitions, rather than a simple unification. Bridge (2017) highlights that even within EE there are different understandings and interpretations of the term, which leads to considerable confusion. Such confusion and lack of distinction can lead to issues of a lack of clarity about course aims and objectives as well as an inappropriate borrowing of course content (Bridge, 2017). If all enterprise and EE are labelled the same, despite potential differences in the course content, learning outcomes, and objectives, there risks a danger of misunderstanding about the course purpose. Similarly, an "entrepreneurship course" could encourage the borrowing of course material and assessment based on the assumption that all courses are similar, which might not always be the case given the different understandings and interpretations of the term. To try and mitigate potential issues around the different understandings of enterprise and entrepreneurship and harmonize how these terms are utilized in higher education, the QAA (2018) guidelines clearly define the terms "enterprise and entrepreneurship", "enterprise education," and "entrepreneurship education".

8.3.2.1 Enterprise and Enterprise Education

The QAA (2018, p. 7) guidelines define "enterprise" as "the generation and application of ideas, which are set within practical situations during a project or undertaking."

This situates enterprise as a generic concept that can be applied to all areas of education and professional life, as the ability to generate and apply ideas in practice is important for all members of the workforce and individuals' personal lives. In order to achieve this, it is identified that a combination of "creativity, originality, initiative, idea generation, design thinking, adaptability and reflexivity with problem identification, problem solving, innovation, expression, communication and practical action" is required. Enterprise education seeks to develop these skills and behaviors and is therefore defined by as "the process of developing students in a manner that provides them with an enhanced capacity to generate ideas, and the behaviors, attributes, and competencies to make them happen" (QAA, 2018, p. 9). To achieve this, enterprise education extends beyond knowledge acquisition, but also includes the development of "emotional, intellectual, social, cultural and practical behaviors, attributes and competences." This gives enterprise education a more practical nature, as it is not solely focused on the transmission of knowledge, but a more holistic development of students to enable them to be enterprising. This holistic development is seen to enhance the students' employability. Enterprise education does this by laying the groundwork of developing students and graduates with an "awareness, mindset and capability to generate original ideas in response to identified needs, opportunities and shortfalls, and the ability to act on them, even if circumstances are changing and ambiguous; in short, having an idea and making it happen" (QAA, 2018, p. 9). Enterprise education seeks to develop a range of enterprise behaviors, attributes, and competences, which are summarized in Table 8.1.

Table 8.1 Enterprise behaviors, attributes, and competencies

Enterprise behaviors	Enterprise attributes	Enterprise competencies
• Taking the initiative • Making things happen • Reflecting • Communicating • Pivoting and adapting • Storytelling • Taking responsibility • Networking • Personal effectiveness • Managed risk taking	• Open mindedness • Proactivity • Curiosity • Self-efficacy • Flexibility • Adaptability • Determination • Resilience	• Intuitive decision making • Identifying opportunities • Creative problem solving • Innovating • Strategic thinking • Design thinking • Negotiation • Communicating • Influencing • Leadership • Financial, business, and digital literacy

Source Adapted from the QAA (2018) guidelines

8.3.2.2 Entrepreneurship and EE

Within the QAA (2018) guidelines, entrepreneurship and EE are identified and defined as distinct from enterprise and enterprise education, with EE building on from enterprise education. Entrepreneurship is defined as "the application of enterprise behaviors, attributes and competencies into the creation of cultural, social or economic value" (QAA, 2018, p. 7). Value can be created by entrepreneurs in a range of settings, including the private, public and third sectors, and such value can be economic, social, cultural, ecological, or emotional (Hindle, 2010). The focus on entrepreneurship as creating value aligns with the perspective that entrepreneurs act on opportunities and ideas to transform them into value for others (Vestergaard et al., 2012). Such value creation can be achieved through the application of entrepreneurs' existing competencies to create something, preferably novel, of value to at least one external stakeholder (Lackéus et al., 2016). It is identified that the creation of value does not necessarily lead exclusively to venture creation, which acknowledges that entrepreneurship can take place in a range of settings and contexts. For example, Pinchot (1985) suggests that entrepreneurs and entrepreneurial-thinking people are crucial within existing firms, as they can think across boundaries and organizational units. Entrepreneurial employees within organizations, often termed "intrapreneurs," have been identified as playing a crucial role in supporting innovation and providing a competitive advantage in established organizations (Blanka, 2019; Guerrero & Peña-Legazkue, 2013).

EE seeks to support students to apply enterprising competencies to create cultural, social, or economic value. "Entrepreneurship Education aims to build upon the enterprising competencies of students who are capable of identifying opportunities and developing ventures, through becoming self-employed, setting up new businesses or developing and growing part of an existing venture" (QAA, 2018, p. 9). To support and achieve this, EE extends the learning environment to focus on students' application of enterprising competencies in a realistic environment. Such a provision allows students to experience entrepreneurship in a practical and hands-on fashion in a safe and scaffolded environment (Bell & Bell, 2020). This allows for the realization of ideas through application in context, thus enhancing students' understanding of business and entrepreneurship processes. It is posited that some overlap exists between the broad set of skills, attributes, and competencies that support graduates' employability, and the characteristics of enterprise and entrepreneurship. Therefore, enterprise and entrepreneurship can help prepare students and graduates for a rewarding professional life. In addition, they are a significant vehicle for delivering the institutional aims of graduate employment, employability, and future success (QAA, 2018).

8.3.2.3 Continued Definition and Conceptualization Challenges

Despite the QAA guidelines providing clear definitions of enterprise and entrepreneurship, and enterprise education, and entrepreneurship education in 2012 as well as a further update in 2018, there remains some challenges and inconsistencies in their application. Jones (2019) suggests that despite the proactive management of definitions by the QAA (2018), enterprise education and EE exist on a continuum. This makes it difficult in some cases to explicitly classify all courses as either enterprise education or EE. Nabi and Holden (2008) suggest that it might be helpful to consider and reconceptualize enterprise education and EE as being on a spectrum, where enterprise education provides the broad and generic skills relevant to most students, whilst EE imparts more specialized skills important for entrepreneurs interested in new venture creation. The skills and competencies located at the enterprise end of the spectrum are particularly suitable for employability, whilst those on the extreme entrepreneurship end of the spectrum, for example, a willingness to take risk, may not be valued by all employers (Sewell & Pool, 2010). It has been suggested that enterprise activities have often been considered as synonymous with entrepreneurship in higher education, with more focused activities aimed at the establishment of new businesses and the less focused targeted at the promotion of more diverse achievements that improve success in the labor market (Moreland, 2006).

Whilst the distinction between enterprise and entrepreneurship might have been implicit within UK higher education, Bridge (2017) suggests that attempts to limit the term "entrepreneurship" to some courses and the term "enterprise" for others often fail as they are only partially done, and the two words are still commonly used interchangeably. The close affinity between enterprise and entrepreneurship has also caused some confusion as to its purpose and remit, as an independent entity from entrepreneurship (Bridge, 2017; Jones & Iredale, 2010). This challenge in identity and distinction, has made defining the contours and boundaries of EE a challenge (Pepin, 2012). Hytti and Kuopusjärvi (2004) suggest that, in practice, EE courses and programs fulfil different roles, depending on their aim and purpose. This highlights the need to not treat entrepreneurship courses and programs as one entity that is comparable across UK higher education, but rather to dig deeper to understand the aim and purpose of individual entrepreneurship courses and programs. As the UK does not have any centrally mandated EE, the QAA guidelines help make a distinction between enterprise and EE, but the distinction is somewhat fluid in practice as HEIs are free to design and develop their own educational provisions. This flexibility allows individual institutions and educators to design programs that they think will be most effective and popular in their specific contexts.

8.3.3 The Delivery of EE

The QAA (2018) guidelines encourage individual educators to decide on the delivery and pedagogy underpinning EE based on the behaviors, attributes, and competencies that they seek to develop. This approach allows flexibility and for educators to take into account students' prior knowledge, experience, and learning. The QAA (2018) guidelines also have a series of suggestions and recommendations to maximize the learning experience. It is suggested that students should have the opportunity to engage with enterprise and entrepreneurship within their own subject and discipline in order to develop entrepreneurial competencies relating to their own field and specialization. To achieve this, the delivery of EE should encourage students to "do" their subject in an entrepreneurial manner, allowing them to identify and solve problems with the goal of developing value propositions for others. To support this, EE can be embedded into various disciplines or take a cross-disciplinary approach where students from different fields of study can be brought together. For example, business, computing, and creative arts students can work together as a team to pool their expertise, which provides the experience of cross-disciplinary working. When students can create value in their own contexts, this allows them to apply their individual unique skillsets and capabilities to their areas of interest, which provides a more tailored learning experience. This has the potential to help students understand what entrepreneurs do in their field and how they can be entrepreneurial in their contexts (Bell, 2020a). Aldianto et al. (2018) argue that it is vital for students to understand value creation and the value creation process, while Lackéus (2018) suggests that learning and teaching approaches focusing on value creation are gaining traction.

To support the development of enterprising and entrepreneurial skills and capabilities, learning activities that are realistic, relevant, highly engaging, and impactful should be utilized. Experiential learning strategies, simulations, and venture programs that link theory and practice, along with the encouragement of reflection after engaging in such activities, are identified as useful means to achieve this. Previous research has suggested that experiential learning is particularly efficacious in EE (Fuchs et al., 2008; Honig, 2004). Experiential learning approaches have been increasingly adopted in UK business schools to supplement traditional teaching (Piercy, 2013). As such, there is a shift away from a purely didactic process-driven approach to teaching entrepreneurship, which is focused on knowledge transmission, to an approach that emphasizes learning from experiences. Experiential approaches to EE offer more opportunities for students to develop the entrepreneurial skills, competencies and mindsets required for entrepreneurship (Bell & Bell, 2020). Macht and Ball (2016) highlight the value of authentic entrepreneurial learning, where students engage in learning activities that provide an authentic experience, imitating what entrepreneurs do in the field. Therefore, whilst experiential learning can take many forms, experiences that are more authentic can promote deeper learning and engagement, and effectively prepare students. Indeed, Ramsgaard and Christensen (2018) argue that working with realistic and authentic projects is vital in EE. To support students' direct engagement with entrepreneurship, the QAA (2018) guidelines suggest that students should be enabled to start and run a business as an integral part of their program. This can be achieved through offering opportunities for students to run small-scale businesses in a scaffolded manner, or through larger-scale venture creation programs. In venture creation programs, the creation and development of an active real-life venture is the primary vessel for learning (Lackéus & Middleton, 2015). The authentic real-life consequences of creating a venture as part of the learning experience can be an emotional ride for students. The challenging and rewarding nature of the experience can support the development of entrepreneurial competencies (Barr et al., 2009; Lackéus, 2014; Meyer et al., 2011).

The QAA (2018) guidelines emphasize the value of engagement with industry and the local community through the delivery of EE. Providing action-based practical activities and challenges set by the community, businesses, and other stakeholders is suggested as an effective way to engage students in enterprise and entrepreneurial activities. This can be achieved through "live" cases studies where local businesses or social enterprises provide briefs for tasks, for example, consultancy, design, or research services. Such engagement with local businesses and stakeholders can provide authentic enterprising and entrepreneurial experiences, allowing students to create value for the community around them. Bringing entrepreneurs, business specialists, and former alumni into the classroom and learning environment is also promoted to link students with industry and the local community. This can be achieved through guest lectures, mentoring, and providing guidance and feedback (Bell & Bell, 2016b). Ensuring the link between practice and theory, and supporting reflection, is important when bringing external expertise into the classroom to maximize learning. Developing links between universities and industry to support meaningful exchange has long been understood to be mutually beneficial (Katz, 1991). It has been suggested that networks provide an essential link between an entrepreneurial idea and successful business creation for nascent entrepreneurs (Adler & Kwon, 2002; Blundel, 2002). In addition to networks, hearing about other peoples' experiences and mentoring can also aid and support nascent entrepreneurs to overcome potential problems and challenges (Ragins et al., 2000; Sullivan, 2000). The sharing of experiences and mentoring can take many forms. This can potentially be delivered by entrepreneurs, business, legal and finance specialists, faculty, and other outside experts, depending on their experience and background (Bell & Bell, 2016b).

Finally, the QAA (2018) guidelines recommend making a distinction between learning *about* entrepreneurship, learning *for* entrepreneurship, and learning *through* entrepreneurship. Making these distinctions can help clarify the purpose of the course and whether it is more focused on the theoretical or practical components of EE. The next section discusses in detail the value of differentiating between the three approaches to the delivery of EE, and their differences and underpinnings.

8.3.4 "About," "For," and "Through" EE

The QAA (2018) guidelines identify and distinguish three ways in which enterprise and EE can be delivered. Each of these three types of EE targets a different type of learning. They encompass a wide range of educational goals and objectives (Bell, 2021), rooted in contrasting and conflicting beliefs as to its value and purpose (Hannon, 2006). Therefore, understanding the distinction between these three forms of EE delivery can help inform learning and assessment strategies, as each type of delivery should be aligned with relevant educational goals and in turn appropriate assessment to ensure constructive alignment. The learning outcomes, teaching content, and learning assessment should all be constructively aligned (Biggs, 2012). This ensures that what is taught and how it is taught can achieve the learning outcomes and goals, and in turn the assessment of learning can measure the learning outcomes. Ensuring effective constructive alignment can motivate students to engage with the teaching and learning, as they can understand how doing so will help them achieve the learning objectives and do well in the assessment. This has been found to be an important part of entrepreneurship curriculum design, particularly when educators seek practical engagement from students (Morselli, 2018).

It has been identified that entrepreneurial knowledge, skills and ability, and experience all play a role in entrepreneurial learning (Bell, 2021; Bell & Bell, 2020), and the three types of EE delivery target these areas. "About" EE focuses on imparting knowledge and theory underpinning entrepreneurship; "for" EE focuses on learning how to be entrepreneurial by developing entrepreneurial skills and ability; and "through" EE emphasizes on providing experience through the application of practical entrepreneurial activity. Whilst all three approaches to EE are valid, the distinction is important when the program or curriculum is being evaluated and new programs are being developed.

"About" EE aims to provide students with an understanding of the theory and knowledge that underpins entrepreneurship. It emphasizes the accumulation of knowledge about entrepreneurship and the entrepreneurship process (Kakouris & Liargovas, 2021). Such teaching can explore a range of topics and themes such as venture creation, business growth strategies, innovation, and social enterprise. This type of EE normally draws upon more traditional pedagogy, such as lectures, set texts, and discussions, to support learning, assimilation, and reflection. As the delivery of "about" EE is focused on teaching the theory behind entrepreneurship, assessment of students' understanding of the theory is often appropriate. For example, assessment through exams or essays, where students can demonstrate what they know about entrepreneurship and its theory, should be aligned with the learning objectives.

"For" EE seeks to develop learners' entrepreneurial skills and competencies to prepare them for potential entrepreneurial endeavors. It provides insight into what it means to be enterprising and being an entrepreneur. In order to develop practical skills and competencies, the delivery of such teaching normally moves away from a focus on traditional didactic teaching methods toward engaging students through active learning. In such an approach, students participate in activities and experiences where they can develop their enterprising and entrepreneurial skills and competencies in a meaningful and relevant context. Such activities and experiences should challenge students to visualize opportunities through the application of creativity and innovation. Students are also encouraged to develop business plans, and then pitch and present their ideas to develop their presentation and persuasion skills (Kakouris & Liargovas, 2021). As learning "for" entrepreneurship is focused on developing practical, enterprising, and entrepreneurial skills and competencies, the requisite learning objectives and assessment should be constructively aligned. Assessment should focus on determining whether students have developed the entrepreneurial skills sought in the learning objectives. To assess this type of learning, practical activities where students can demonstrate their development are required. Practical forms of assessment are more efficacious in determining students' practical skills, so assessment "for" entrepreneurship typically involves presentations or coursework, rather than the testing of knowledge and understanding through tests or exams. The development of business models and pitches relating to the students' own contexts and aspirations can be effective in supporting them to become entrepreneurial. At the same time, it provides insight into the development of new business and potential entrepreneurial opportunities. To mimic the real entrepreneurial process, groupwork and collaboration can be used to develop effective leadership, responsibility, teamwork, and coordination skills. The development of such skills will also support employability. The learning in "for" EE can take many forms and levels or practicality, ranging from the simple generation of ideas to the enactment of some parts of the entrepreneurial process, such as pitching, planning, or selling. This means that there is potentially some crossover between "for" and "through" EE, as both involve undertaking parts of the entrepreneurial process. However, "through" EE represents a more holistic engagement with entrepreneurship and the entrepreneurial process.

Teaching "through" EE seeks to develop entrepreneurial capabilities by providing students with the experience of engaging in entrepreneurship. This gives students the opportunity to draw on and apply their knowledge of entrepreneurship as well as their enterprising and entrepreneurial skills and competencies. This allows the exploration and transformation of knowledge through experience (Bell & Bell, 2020; Kakouris & Liargovas, 2021). In this type of teaching, students learn from their experiences and should be encouraged to reflect on them. They may experience entrepreneurship in groups or individually, depending on the type of course and what is most appropriate in the learning environment. Working in groups supports the development of important employability skills, and delegation within a group provides opportunities to divide and share tasks and responsibilities. Learning "through" entrepreneurship is often supported by dedicated facilitators or mentors who can sit alongside business incubation and accelerator schemes. This potentially allows for "through" EE to be a gateway to entrepreneurship. As the focus of "through" EE is learning from experience, it is primarily a reflective process where after engaging in entrepreneurship or entrepreneurial activities, students map their own learning and progression. As a result, assessment is commonly in the form of reflections on business decisions including pivots or iterations and business outcomes. Failure forms part of the entrepreneurial learning process, and it can be used to support the learning and development of resilience (Kauppinen et al., 2019). Assessment may not focus solely on entrepreneurial success, but rather how challenging situations were, or could be, addressed and the business pivots made. Reflecting in this way supports the realization of failure and learning from it.

The concepts of "about," "for," and "through" EE build on one another and can be supplementary. As is common with most educational disciplines, practice should be underpinned by theory. Therefore, before students move onto developing entrepreneurial skills and competencies and engaging in entrepreneurial practice, they should understand the knowledge and theory that underpin entrepreneurship (Bell & Bell, 2020). This can be achieved by blending the delivery of knowledge and theory alongside practice within a module, or by structuring EE to first deliver modules focused on theory and then modules focused on building entrepreneurial skills and competencies as well as providing entrepreneurial opportunities.

Bloom's taxonomy has widely been used to guide curriculum design and development in higher education in the UK and other parts of the Western world. As a result, it has become the most widely used framework for instructional design and evaluation (Kakouris & Liargovas, 2021). A common use of Bloom's taxonomy is to classify curriculum objectives and assess them to ensure the breadth (or lack of) of objectives across the six categories in the taxonomy (Amer, 2006). The taxonomy provides definitions for each of the six major categories in the cognitive domain. The categories are ordered from the simplest to the more complex and from the concrete to the more abstract (Krathwohl, 2002). The taxonomy represents a "cumulative hierarchy" in which each previous simpler category is a prerequisite to the mastery of the next, more complex category (Krietzer & Madaus, 1994). The six levels in the hierarchy represent essential skills for students to become critical thinkers (Murphy, 2007). For many, Bloom's taxonomy serves as a basis for the so-called "higher order" thinking skills. The taxonomy was revised and updated in 2001 (Anderson et al., 2001) to provide a more dynamic classification, using verbs rather than nouns. These revised levels within the taxonomy going from the most basic are: remember, understand, apply, analyze, evaluate, and create. They encourage movement beyond knowledge comprehension and toward the higher-order skills of analysis, evaluation, and synthesis (creation) (Bloom, 1956). "About" EE focuses on achieving and demonstrating the lower levels of Bloom's taxonomy: remembering and understanding the theory behind EE. Exams and essays can be used to assess students' ability to recall, summarize, and explain the theory. "About" EE can also be used to encourage students to apply theory, concepts, and knowledge to case studies for analysis and coursework. Compared to traditional teaching commonly found in "about" EE, "for" and "through" EE is centered on the core principles of active learning: student activity and engagement in the learning process (Prince, 2004). The active and experiential learning that forms part of the learning "for" and "through" EE requires the higher-order thinking skills of analysis, evaluation, and synthesis (creativity) (Bell, 2015). "For" and "through" EE moves away from knowledge retention to knowledge harvesting, which requires students to synthesize knowledge critically to create something new. When preparing students "for" entrepreneurship, it is important to develop their critical thinking skills and analytical judgement to make entrepreneurial decisions based on the analysis and evaluation of information and context. "For" EE can also encourage students to synthesize information to create business models and plans that can be pitched to an audience. Similarly, "through" EE requires students to undertake analysis and evaluation via the entrepreneurial process and create value within an entrepreneurial venture. Therefore, "for" and "through" EE focus on achieving and demonstrating higher levels of thinking within Bloom's taxonomy of learning. However, as previously mentioned, it is important to provide a grounding to the theory underpinning entrepreneurship, and this aligns with the concept of moving students up through the levels of Bloom's taxonomy (Lord & Baviskar, 2007). This transition moves students through the process of receiving, understanding, and applying entrepreneurship theory, analyzing, and evaluating entrepreneurial decisions and opportunities, and finally creating something new.

8.4 EE Approaches and Characteristics

Different educational philosophies and theory underpin the educational approaches of "about" entrepreneurship, which teaches the theory behind entrepreneurship; "for" entrepreneurship, which develops learners' entrepreneurial skills and competencies so that they are ready for potential entrepreneurial practice; and "through" entrepreneurship, which supports learning by practising entrepreneurship (Bell, 2021; Hannon, 2005). To achieve the different goals of these three types of EE requires different forms of pedagogy. These are underpinned by a range, and potentially a mixture, of educational philosophies and theories. Understanding how different educational philosophies and theories inform and support the type of EE helps inform educators as to the most effective delivery of teaching and assessment of learning (Hannon, 2006). Thus, educational philosophies and theories guide educators through the contrasting landscapes of educational approaches to EE (Ramsgaard, 2018). As the chosen educational philosophy guides teaching and pedagogic practice, it can be a differentiator between educators who are, and those who are not, aware of what they are doing and why (Merriam, 1982). Whilst educators may not always be able to explicitly articulate their philosophical orientation, nevertheless an implicit educational philosophy, or mix of, will still be underpin, direct, and drive their teaching and pedagogic practice (Darkenwald & Merriam, 1982). Educators may not always explicitly choose or identify with a particular educational philosophy or theory of learning to guide their practice and may follow the path of least resistance, or might subconsciously adopt one, but this will still impact how they teach and assess learning.

Within the delivery of EE, it is possible to mix and integrate several educational philosophies and theories that guide the teaching and assessment. Given the potentially diverse and multifaceted aims of EE and the diverse pedagogic approaches, it is suggested that no single philosophy or learning theory can effectively guide EE (Ramsgaard, 2018). Robinson et al. (2016) suggest that when delivering student-centered EE, it is beneficial to bring together different learning theories, philosophies, and approaches to EE, as this could encourage the development of entrepreneurial awareness. This is supported by Brieger et al. (2020), who posit that there is no single adult learning theory, but many philosophies and theories that need to be interwoven. Therefore, it is beneficial for educators to understand the interlinking nature of the different educational philosophies and theories that inform the learning process and pedagogic practice; doing so can help educators maximize the effectiveness of their teaching (Bell & Bell, 2020).

The behaviorist approach to teaching leads to the transfer of objectivist knowledge and the assessment of students' ability to reproduce the knowledge taught. This approach involves the didactic transfer of knowledge, which leads to the efficient and functional mechanical processes (Löbler, 2006). Repetition, reinforcement, and assessment through testing are the central features in this approach to teaching and learning. It has been argued that a system based solely on providing objectivist knowledge does not support and encourage the creativity of learning (Freire, 2006).

It has also been argued that this passive approach to learning can result in the disengagement of students and superficial learning as students are only expected to repeat the information that they have been taught (Bennet, 2006). In this approach, the quantity and quality of the information and knowledge transferred are paramount, as opposed to the learning experience and the value that they offer (Dierksmeier, 2020). The information and knowledge transferred in this way can lack context and not relate to students' past knowledge and experiences. The educator's role in this approach is to manage, predict, and direct the outcome of the learning; in this way, they can meet the perceived needs and requirements of the students in the curriculum, which should be aligned with societal and industry requirements (Hannon, 2006). This approach, whilst commonplace in EE curriculum, is increasingly viewed as being dated and of value only for the transmission and transfer of theoretical and instructional information (Gedeon, 2014). However, education through the passive delivery of objectivist knowledge may be adopted for expediency, in which case it is not philosophically or pedagogically informed (Bell, 2021). The behaviorist approach based on the didactic tradition of objectivist knowledge has been regarded as a potentially suitable approach for the delivery of "about" entrepreneurship. This approach can be underpinned by the focus and emphasis in "about" EE on the accumulation of knowledge, the theory about entrepreneurship, and the entrepreneurship process. However, the delivery of "about" EE using only a behaviorist approach may limit the depth of learning, with the learning of knowledge and theory not translating into understanding and application. Whilst "for" and "through" EE seek to develop more practical outcomes in terms of entrepreneurial skills, competencies, and capabilities, it is important that theory underpins practice. The behaviorist approach can inform students of the underpinning theory and frameworks that they can then use to analyze their own experiences (Peltier & Scovotti, 2010). In addition, the didactic approaches adopted within behaviorist teaching can be effective at providing instructional information for the practical learning activities and experiences included in "for" and "through" EE. In more practical and hands-on teaching, there is still a need for students to understand the learning objectives, what they must do, and how they will be assessed (Béchard & Toulouse, 1998; Bell & Bell, 2020). Therefore, the behaviorist learning approach can act as a prerequisite stage for "for" and "through" EE.

Cognitivism moves away from behaviorism in that it involves the processing of knowledge rather than just the memorization of it (Brieger et al., 2020). In this approach, the student plays a more active role in the process of learning. The cognitivist paradigm postulates that students can develop objective knowledge to reach new understanding through reasoning or intellectual intuition (Kyrö, 2015). In this case, learning is an internal process that involves memorization, thinking, reflecting, and abstraction of the information (Ally, 2004). Cognitivism is thus based on the individual's cognitive processes and the level of an individual's cognitive development (Bandura, 1977). Learning is the product of the student's critical thinking to process the information to create new knowledge (Bell, 2021). The cognitive approach focuses on the student's development by changing the way they think and view information. Within such learning the educator's role is to promote and facilitate learning, rather than directing it (Hannon, 2006). Cognitivist teaching approaches can include the use of case studies where students apply knowledge and theory to critically analyze a particular case. Learning from case studies has been highlighted as a commonly utilized approach that is grounded within cognitivism, and which allows students to develop critical insights that can then promote theoretical understanding (Kantar, 2013). Such activities can promote the development of critical thinking skills through the application of knowledge and theory to a particular context. The cognitive paradigm drawn from rationalism posits that objectivist knowledge gained through reasoning or intellectual intuition, once taught and understood, can then be transferred into other situations. The cognitivist approach is particularly efficacious in supporting the development and nurturing of critical thinking and reasoning skills. This makes the cognitivist approach suitable for "about" EE as it supports moving beyond remembering knowledge to furthering students' understanding as well as supporting application to context. Through application to context, knowledge and theory are critically applied to support analysis and evaluation. This makes cognitivism a critical and essential tool for inquiry (Facione, 1990). Like behaviorism, cognitivism can be supplementary and a prerequisite for both "for" and "through" EE. Entrepreneurship requires the development of reasoning and critical thinking skills that can support "thinking that is purposeful, reasoned and goal directed—the kind of thinking involved in solving problems, formulating inferences, calculating likelihood, and making decisions" (Halpern, 1996, p. 5). Cognitivist learning allows the structuring and storing of knowledge and theory, which can be retrieved and applied when required during hands-on learning and the entrepreneurial process. Therefore, cognitivism can be a complementary part of "for" and "through" EE. Cognitivism, however, does not consider the arguably experiential nature of entrepreneurial learning (Rae, 2005). This ensues from learning by doing and reflecting on the process (Cope, 2005). As a result, the cognitive approach is limited since it does not take into consideration the importance of developing reflexivity and self-awareness (Ferreira, 2020).

Behaviorism and cognitivism are based on the premise that knowledge is an objective phenomenon. This has been increasingly challenged by constructivism and constructivist approaches to learning. As a theory of knowledge constructivism emphasizes the part played by the individual in creating their own individual meaning from knowledge based on their own context and experience (Mueller & Anderson, 2014). Knowledge construction is an active and interpretive process, and therefore based on the subjective understanding of the individual where meaning is derived from past and present knowledge (Merriam et al., 2007). Educators adopting a constructivist approach ground their teaching in active participation, which involves the creation and development of active learning in realistic contexts where students think independently and question their experience in the process of discovery (Mathews, 2007). In the learning process, the role of the educator is to guide students to construct their own meaning and understanding of the world around them, and not to simply mirror the educators teaching (Jonassen, 1991). Constructivism can be used to underpin a wide range of active learning approaches. These can problem-based learning, inquiry learning, and experiential learning (Kirschner et al.,

2006). The construction of meaning from new knowledge created through new experiences and problem-solving can reflect the knowledge creation and decision-making within the dynamic context of entrepreneurship (Bell, 2021). This type of approach to learning prepares learners for entrepreneurship by meeting challenges and problems. To ensure an effective learning process, the educator should design it to be constructively aligned (Biggs, 2012). To ensure this, students should understand the learning process and their role within it, what they are expected to achieve, and how they will be assessed. The educator should ensure that the learning experience is scaffolded, if required, to enable students to engage in the learning process. Reflection is vital in the constructivist learning approach to enable students to create meaning and sense from their experience and its application to the real world. Whilst the constructivist approach is not well aligned with the goals and focus of "about" EE, as it seeks to develop subjective knowledge and understanding in relation to a student's individual context, it can be of value in both "for" and "through" EE. Constructivism underpins a range of hands-on and active learning pedagogic approaches that are effective at developing skills and ability through reflection. As "for" EE seeks to develop entrepreneurial skills and competencies and "through" EE the skills to develop entrepreneurial capabilities, active and experiential learning where students can practice, reflect, and develop their abilities is essential. One distinction between the learning in "for" and "through" EE is the nature and context of the activity or experience. The learning experience within a "for" EE approach will generally be less focused on actual entrepreneurship and new venture creation, but will instead encourage students to engage in the entrepreneurial process so as to provide them with insight into how to be enterprising and being an entrepreneur. A "through" EE experience will allow students to engage in the full entrepreneurship process and experience to support the development of their entrepreneurial capabilities.

As EE moves away from the didactic transmission of knowledge to developing more practical entrepreneurial skills and competencies, the roles of the teacher and student in the educational process changes. As learning shifts from behaviorist, cognitivist, and constructivist approaches, the emphasis moves from the teacher to the student. The constructivist approach is student-centered: students take the lead, and take responsibility for their learning and be self-directed, while the teacher acts as a guide or facilitator (Seikkula-Leino et al., 2010). This is in contrast to the behaviorist approach where the teacher acts as the purveyor of knowledge and takes the lead in the learning process. It is increasingly recognized that students, rather than teachers, are the primary agents in the EE process (Aparicio et al., 2019). However, teacher-centered approaches are particularly effective when supporting students' theoretical understanding of entrepreneurship and its benefits compared to student-centered approaches (Hytti & O'Gorman, 2004). Ismail et al. (2018) highlight and discuss the importance of context and culture on the effectiveness of teacher-centered and student-centered approaches. Where society widely accepts the authority and wisdom of teachers, teacher-centered approaches may have a greater impact on the development of entrepreneurship intention. In addition, students who are unfamiliar with constructivist student-centered approaches may face challenges when taught using such approaches (Bell, 2020b). These challenges include understanding the learning process and how learning and knowledge are discovered and created, linking taught content and knowledge to practical activities, working in, and managing group interactions (Bell, 2020b). Harima et al. (2021) also found that role ambiguity in the learning process may exist for students in courses centered on venture creation. Therefore, it is important that students fully understand the pedagogic and learning process within student-centered EE and the expected role of the student.

The next section discusses two case studies of EE practice in the UK and links back to the different approaches to EE and the underpinning educational philosophies and theory. The first case adopted a progressive "for" EE approach within the curriculum, whilst the second case study adopted a "through" EE approach within an extracurricular activity.

8.5 Examples of EE Practice in the UK

This section provides two case studies of EE in the UK. These case studies will be used to apply the concepts from the QAA guidelines and the educational philosophy and theory discussed in the previous sections. These case studies do not necessarily represent best practices, and there is a need to tailor the learning experience to individual student cohorts and contexts. The two case studies both apply experiential learning approaches, as it has been suggested that such approaches are particularly efficacious for EE (Fuchs et al., 2008; Honig, 2004). Experiential learning allows students to actively participate in the entrepreneurial process, rather than simply reading or hearing about it (Jones & English, 2004). The learning within both cases studies sought to mimic part of the entrepreneurial processes and the creation of something, using new processes or techniques to create value (Schumpeter, 1942). This allowed for an inductive process of opportunity recognition in an environment with a degree of uncertainty, unpredictability, and risk (Jack & Anderson, 1999). This meant that the students needed to engage in creative problem-solving throughout the experience, and learned by doing (Jones & Iredale, 2010). Due to the acknowledgment of uncertainty and unpredictability in the students' activities and tasks, the teaching within the case studies moved away from a process view of entrepreneurship whereby it was viewed and conceptualized as a set of linear stages, to a method view that allowed students to engage in the method of entrepreneurship (Neck et al., 2014). Adopting a method view of entrepreneurship forces students to go beyond seeking understanding, knowing, and discussing entrepreneurship to applying and acting through practice (Neck et al., 2014). The learning activities within the case studies allowed students to play freely with only limited burden in an entrepreneurial setting, develop empathy with those they were developing products for, and create and experiment within entrepreneurship (Neck et al., 2014). The activities also provided links to real-life entrepreneurial practice in a scaffolded environment to encourage and support the learning and development of entrepreneurial skills (Macht & Ball, 2016).

Within both case studies, the underpinning theory on entrepreneurship was provided to students, as engaging in entrepreneurship does not exclude theory; rather, the effective undertaking of entrepreneurship requires a set of practices that is grounded and underpinned by theory (Bell & Bell, 2020; Neck et al., 2014). Therefore, there were different teaching goals within the cases studies at different stages, and to effectively achieve these different goals, a mix of educational philosophy and theory was used. The main emphasis was on learning from experience: reflecting on the experience and what worked well and did not work well, and what changes could be made. This was encouraged and supported by Schön's (1983) principles of reflection-in-action, which supports reflection during the experience to re-evaluate logic and strategy, and Kolb's (1984) theory of experiential learning, which promotes turning experience into learning and development after the event (Bell & Bell, 2020). The overall purpose within the teaching was to support students to acquire entrepreneurial skills, knowledge, and mindset through deliberate hands-on, action-based activities that enhance and support the development of entrepreneurial performance (Neck et al., 2014).

8.5.1 Case Study 1: Popup Shops and Social Events

Course Purpose and Structure

This case study presents details of a course where students engaged with the entrepreneurial process by undertaking two practical applied tasks within groups. Each of the applied tasks was targeted at part of the entrepreneurial process, rather than the whole entrepreneurial process and the formation of a complete business. As such, the course could be classified as a progressive "for" EE approach. The course was designed to give students the experience of identifying entrepreneurial opportunities, planning, managing the sales and marketing process, and implementing the initial launch of a venture in a supportive and scaffolded environment. The tasks and learning environment were designed to be fun and positive. Students could play freely with ideas, connect with end users to develop empathy and understand their needs and desires, as well as create, experiment, and reflect on how the process went (Neck et al., 2014). The course was for second-year higher education students who already had an understanding of basic business functions and the supporting theory. It allowed students to apply some of the theory previously taught to them, as well as new theory and knowledge delivered as part of this course, to practice and engage in the entrepreneurial process. Whilst the course was mandatory for some students, it was an elective for others. The course ran throughout the full academic year, a total of 24 weeks, with 12 weeks in each semester. Midway through the first semester, students needed to complete the first applied task, whilst toward the end of the second semester, they needed to complete the second applied task.

Applied Task 1: Running a Popup Shop

Students in groups were tasked with picking a product that they would then attempt to sell in a popup shop. They were presented with a booklet detailing the products they could choose from. The booklet contained details about the products and their wholesale cost. The products were largely imported and represented trends in other countries. The groups had the opportunity to review the potential products and costs, and then decide which product they wanted to sell, along with the price they would charge, and how they would market it. They created a list of the products they would like to sell in their preferred order. The groups were encouraged to develop a robust rationale and justification for their choice of products. One of the rules was that no two groups could sell the same product. So, if more than one group requested the same product, there would be a tiebreaker to decide who could sell the product. The tiebreaker involved the groups pitching their rationale for choosing the product and why they would be the best representative for it. The decision as to who got to sell the product was made by the instructor in consultation with the rest of the class. The competition and experience of competitive pitching was meant to create a buzz. This component of the class and the selling of the product through a popup shop emulated a task regularly undertaken on the reality-television show *The Apprentice* and parallels were drawn between the two to heighten the similarity and enjoyment. Once it was confirmed what each group was going to sell at the popup shop, students were required to confirm their pricing strategy, marketing, and advertising, and complete a worksheet explaining their expectations and predicted sales. In the following week, the pop-up shop was run by a second different group of students selling their using their marketing and advertising materials developed and the pricing strategy set. During and after the popup shop, students were asked to reflect on the experience and the worksheet previously completed, with their expectations used as an aid for their reflections.

Applied Task 2: Developing, Organizing, and Running a Social Event

As part of the second half of the course, the student groups needed to develop an innovative and novel concept for a social event, pitch it to the Student Union, and then once agreed after making any required changes, market, organize, and manage the event. This task built on the first by allowing students to engage with more of the entrepreneurial process and involved more complexity. Each group had an evening where they could run an event at the Student Union on campus. They developed a theme and concept for their event, including the entertainment and activities to be offered. Additionally, they could pick the entry price and drink offers from an agreed range, and they needed to market the event. In essence, students acted as the event promoter, whilst the regular staff managed the essential functions of service and security, etc. Each group was provided with a budget that they could spend to cover props, activities, and entertainment for their event, and the entry/door cover charge and a percentage of bar sales formed the revenue from which their final profit was determined. Each group was tasked with breaking even or making a small surplus.

As part of the preparation for the second applied task, students were guided and supported through several stages of the entrepreneurial process including market research, planning, budgeting and costing, and marketing and advertising. They needed to complete a document like a business plan for their event, including details of their expectations for the event. During the process, students were encouraged to reflect on how the process was going as well as on the outcome and success and/or failure.

The students ran a range of events with different themes, activities, and entertainment, including trivia, karaoke, food events, sporting activities, photobooths, bands, and DJs. Some events failed to break even, while others made a surplus.

Teaching During the Course

Due to the hands-on and practical nature of the course, it was timetabled for three hours per week in a single block for 24 weeks. All the teaching was delivered in seminar rooms where the students sat in their respective groups. The first session introduced the course and provided an overview of the module, outlining the learning objectives, the scope of content, what students would need to do and their role in the learning process, and how they would be assessed. After this, each session focused on what the students needed to do to prepare for the applied tasks. The instructor used approximately the first third of each session to introduce the week's topic, discuss the theory underpinning the topic, and explain what students were required to do. During this first phase of the session, the instructor provided theory and knowledge and encouraged critical analysis and discussion. The presentation and discussion of objectivist knowledge were underpinned by a behaviorist and cognitivist approach to teaching and learning. After this, the students worked in their small groups to apply the theory and knowledge to the activities and challenges to prepare them for the tasks. For the final two-thirds of the session, the activities and challenges were the same for the whole class, as each group was working at their own pace and doing projects with different emphases based on their own assessment of their context and situation. Therefore, there was some flexibility allowing groups to focus on and prioritize what they thought was most important for them. During this phase of the session, the instructor acted as a facilitator and guide who was available to explain what should be done and why, and provide guidance when groups were stuck. This allowed students to construct their own learning and meaning based on their own context of what they were doing, how they approached it, and how it went. Therefore, there was limited uniformity of the learning from this phase. Time for structured reflection was built into the course and individual groups and students were encouraged to reflect on the decisions made each week, how it was going, and whether any pivots or adjustments should be made. Additionally, students were encouraged to reflect on how the applied activities had played out and what they would do differently next time. Both reflection-in-action and reflection at the end of sessions and after the applied tasks, were central to learning throughout the course (Kolb, 1984; Schön, 1983). The learning during this phase of the sessions was underpinned by a constructivist educational approach. As the course progressed, students were generally given greater autonomy and less guidance as they became more accustomed to what they needed to do, took more responsibility for their learning, and developed their entrepreneurial skills. The groups became experts in their own events as they continued to research and plan them. The theory and the stages for completing the activities were presented in bite-size pieces, which supported them to achieve what many of them had initially gasped at when they found out what they needed to do.

Assessment

The assessment was a portfolio comprising three parts that had been completed at different stages of the module. Feedback was given after the submission of each component for students to improve on their subsequent submissions. The first component required students to present a group reflection of their experience running the popup shop (applied task 1). Help was provided to structure the reflective presentation, as this was many students' first reflective presentation. They were encouraged to revisit their expectations and forecasts, choice of product, marketing, pricing, what worked and did not work, and why this might have been. They also identified what they had learnt by exploring whether there was anything they would have done differently and what advice they would give to someone considering opening a retail shop. For the second component, the groups wrote a report explaining and justifying their decisions made for the social event (applied task 2), which was submitted before the event ran. This covered the concept for the event, activities, and entertainment, the marketing strategy and messages communicated, their financial and budget decisions as well as any other decisions they had made. The third and final assignment was an individual reflection on the social event (based on applied task 2), a small reflection on the entire module, and what was learnt. Like the first assignment, students were guided to reflect on the experience and their learning. As students moved through the module, their reflective capability and ability to learn from their experience were strengthened.

Working with Partners and Stakeholders

As part of the course, students had the opportunity to interact and work with a range of stakeholders. They talked to professionals about their entrepreneurial ideas to gain their buy-in. As part of the popup shop, students discussed their pricing and marketing plans to seek approval from the course instructor and a representative of the property management company who ran the shopping center where the retail unit was located. Also, once the popup shop was up and running, the students had to speak with customers. Whilst preparing for the social event, students needed to undertake market research to determine the likely success and viability of their concept. They also discussed and cleared their event concept and plans with the Student Union and bar manager. These interactions with external stakeholders helped create a real-life experience and developed students' persuasive communication and soft skills. In addition, students had the opportunity to build networks and entrepreneurial identity through holding entrepreneurial discussions.

Student Reaction and Learning

The student reaction to the module was largely positive as they enjoyed the experience and it stood out from the other courses they took. The module helped them link and apply to actual practice the theory taught in the course and other business courses that they had previously studied. However, some of the students found it quite intense as it was not a class where they could be passive. This meant that whilst offering the potential for a real sense of achievement, it was time-consuming and tiring. This was highlighted and discussed when students were choosing the class, so as to ensure that those who enrolled were aware of the expectations and could commit to it.

The general learning that often results from the course includes the importance of linking products and concepts to customer demand and interest in order to identify a successful opportunity. Groups regularly got sucked into choosing products that they liked for the popup shop, and developing concepts, activities, and entertainment for the social event which they thought would be enjoyable and popular. Instead, they ought to have considered what the market and potential customers would be interested in. This also highlights the importance and value of market research and not taking the easy option of speaking to people within one's social group as others may give a more honest and unbiased view. The importance of critical analysis and evaluation in decision-making was regularly brought up in reflections, as groups regularly struggled to make effective decisions. Groups could either become paralyzed by different viewpoints or acquiesce to one person for all the decisions. Presentation and the importance of being explicit, concise, and persuasive often comes to the fore when pitching and speaking to external stakeholder to seek agreement. External stakeholders often have defined red lines and questions that groups might not be able to answer on the spot. Groups also often learned the difficulty in developing effective budgets and forecasts, and regularly overestimate sales while benchmarking the value of their product. Individuals regularly reported feeling more comfortable with the concept of entrepreneurship and knowing where to start if they were interested in starting a new venture. Similarly, the engagement with external stakeholders made them feel more confident that they would be taken seriously when approaching such conversations. Finally, students often reported they learned the value of reflection and how to learn from it, as this was an important step when dealing with the many issues that often occurred, both when planning and during the applied tasks.

Further Information

This case study outlines and discusses an updated and refined version of the course analyzed in Bell (2015). For further critical analysis and details as to the effectiveness of the original course in developing students' entrepreneurial skills, see Bell (2015). The original course was kindly supported by the Academy of Marketing, Teaching and Research Development Grant, and the nature of the course allowed for it to be sustainably funded, with any profits/surpluses used to support future academic years.

8.5.2 *Case Study 2: Supporting New Retail Ventures*

Extracurricular Activity Purpose

This case study details an extracurricular EE program. It has been identified that, in some contexts, extracurricular EE has more scope for creative and innovative teaching, as there are fewer formal requirements, expectations, and restraints compared to EE delivered in the curriculum (Cui et al., 2021). Involving the collaboration and support from a bank and a property management company, the program provided an opportunity for students to pitch for funding and free retail space in a shopping center. The property manager donated a shop unit in a shopping center while the bank provided some initial capital for four startups. Both were also involved in the training and mentoring of students. The program was promoted as a scaffolded, supported way to try out a retail concept where students could use their summer to gain entrepreneurship experience, market exposure, and potentially make a profit by running a startup. The program could be categorized as "through" EE, as the students were able to run their new venture over a sustained amount of time and complete the full entrepreneurial process.

The program was open to all students at one HEI, regardless of discipline and age. Students had the option to work as a group or independently, depending on their preference. In total, over 50 students participated in the program initially, with the number falling as the program progressed. This attrition was expected as the program grew increasingly demanding and real, and students reassessed their interest in entrepreneurship. Nonetheless, even those who started the program but did not complete it would have received exposure to entrepreneurship and learnt valuable lessons about entrepreneurship and themselves.

Stages of the Program

The program was broken down into three stages. The first stage involved mentoring from a range of consultant mentors including a retail expert provided by the property manager, small-business advisers from the bank, and the entrepreneurship faculty from the institution. Several short sessions were delivered which included a range of topics such as idea generation and refinement, planning, budgeting, persuasive communication, and pitching. These sessions were designed to support students to develop their new venture propositions and write a small business plan, based on the banks' normal loan application. At the end of the first stage, students who wanted to move to the next stage needed to support their business plan, with all business plans receiving feedback. Eight business plans were chosen to move to the second stage where the students were provided with individual mentoring and support to further develop their ideas and plans. They would pitch their plans to a panel who determined the awarding of capital and retail space. Students had approximately one month to develop their plans and pitches. Of the eight groups who pitched four were provided with seed capital and part of a retail unit to run their operations over the summer break. The third stage of the program involved the actual running of the business over the summer. During this time, students had weekly meetings with their mentors and were encouraged to reflect on what was working and not working, as well as what adjustments could be made. This led to several of the businesses making marketing and/or pricing changes. The program largely adopted a constructivist approach to teaching where students were in control of their own learning, and it was structured around the development of their own proposed and actual businesses. Students were supported by industry experts and faculty who acted as mentors and guides who helped them develop their own unique business concepts. Therefore, the learning experience was different for each student. The program went beyond a common business pitching competition: those who were successful received the resources to run their new ventures over the summer and were supported by mentoring throughout the whole process, from the initial idea generation through to completing the paperwork to close or make the business dormant.

Student Learning and Program Outcomes

Student participants stated that the program demystified the entrepreneurial and startup process and what support was available. They also indicated that they now felt more comfortable talking and communicating with those in the industry. Those who ran businesses over the summer highlighted that they had learnt a lot from the experience. This included the need to be adaptable and flexible when their assumptions and expectations did not hold true, for example, items were often priced too high and/or potential customers did not perceive the value of the products as expected. The mentoring also allowed students to build industry networks when they engaged with stakeholders in the industry. Students identified the value of mentors in helping them think through and reassess their decisions as well as reconnect their experiences to the initial decision-making process. This allowed them to review and reflect on their assumptions and why they might not be right. Students also highlighted the value of reflecting during the experience as it allowed them to experiment and make ongoing decisions to improve their business.

Further Information

This case study outlines and discusses a further iteration of the program analyzed in Bell and Bell (2016a).

8.6 Challenges and Future Directions for EE in the UK

This chapter discusses the development of EE in the UK, which continues to evolve despite its relatively long history. Whilst its relevance and pertinence have been maintained in government policy, the way in which EE has been viewed and delivered has continued to change. For example, there is increasing emphasis on social entrepreneurship, and the notion of entrepreneurship has been broadened beyond new venture formation to include value creation in terms of cultural, social, or economic value. This movement to consider entrepreneurship as not exclusively new venture formation is likely to continue to grow, but it may also complicate the definition of entrepreneurship and therefore what exactly should be taught. Some scholars such as Bridge (2017) already believe that making a distinction between enterprise and entrepreneurship is arbitrary and causes more confusion. However, opening up entrepreneurship and EE to more people through a revised conceptual understanding should be considered to be a positive step.

Whilst EE is still taught using didactic principles in some cases, it is increasingly being taught with an emphasis on the development of practical skills and experiential learning. The QAA (2018) offers a set of guidelines rather than a prescriptive code or requirement, which allows entrepreneurship educators to be creative and tailor their provisions to individual cohorts. One of the main points of the QAA (2018) guidelines encourages educators to consider the purpose of their teaching—is it to teach "about," "for," or "through" entrepreneurship? Once this has been determined, there is value in considering and reflecting on the philosophy and theory that effectively underpin this type of EE. A challenge still potentially exists in supporting both students and educators to make the transition between "about," "for," and "through" EE. To support the transition, students need to understand the difference in their role in the learning process and be prepared to take ownership of their learning. Students commonly also need increased scaffolding to start with, which is then removed, otherwise they may become lost or dependent on the instructor. Both scenarios have the potential to reduce the effectiveness of the learning experience. Educators may be hesitant to transition as they perceive that "for" and "through" EE are more challenging, time-consuming, and resource-intensive, and this may not always be recognized by institutions (Bell & Liu, 2019). Therefore, educators may become apathetic and follow the path of least resistance, not wanting to make changes and sticking with the way courses have always been taught. Conversely, there is a potential danger of educators becoming carried away with more progressive EE at the expense of theory. One of the benefits of ensuring that EE is underpinned by educational philosophy and theory is that it reminds us of the importance of underpinning entrepreneurial experiences and hands-on learning with theory and that the two should not be separated. Therefore, there is a risk of educators perceiving their class to be either about theory or practice, but both should be brought together to ensure effective learning.

As technology opens up access to more markets and opportunities, and potentially making entrepreneurship more achievable, it is important that EE and educators can integrate this into the curriculum. Therefore, it is important for technology to be embedded within EE, where students are equipped with the necessary understanding and skills to take advantage of changing and developing technology such as coding, web design, and crowdfunding, etc. This requires educators to stay on top of the latest technology and innovations and to bring them into the classroom, so students can learn, explore, and experiment with them. Similarly, it is important that entrepreneurship continues to be embedded within other disciplines and is not taught in isolation, as it is by allowing students to be entrepreneurial within their own context and using their own unique knowledge base and skill sets that the strongest innovations and opportunities will emerge. The so called fourth industrial revolution is not mentioned explicitly in the QAA (2018) guidelines, however the guidelines highlight that EE should prepare students for job roles which do not exist yet and technologies which are yet to be invented. The fourth industrial revolution has the potential to reshape markets and open new opportunities for entrepreneurs through the embedding of new technologies and e-commence (Hassan et al., 2020). Therefore, EE needs to reflect these new potential opportunities by ensuring that future entrepreneurs are ready to take advantage of them (Bell & Bell, 2023). This can be achieved by ensuring that technological and digital disciplines feed into EE, so that entrepreneurship is not seen as an isolated discipline. Students need to be able to understand how contemporary technology can be used to support opportunity identification and take full advantage of opportunities within the changing landscape. However, on the ground different departments and expertise can become siloed, which can make bringing the required expertise into the EE classroom a challenge. Future EE guidelines might benefit from more explicit guidance as to the role and importance of embedding technology within EE teaching, and by suggesting paths forward as to how this can be achieved in practice.

Effective entrepreneurial ecosystems play an important role in the transfer of students from EE to entrepreneurship. Conducive entrepreneurial ecosystems can complement EE and support students to start new ventures. An effective and strong ecosystem has been described as bringing together alumni, partners in industry and commerce, joint research projects and incubators to offer opportunities for providing encouragement, and the practice and development of entrepreneurial ideas (Miller & Acs, 2017). However, entrepreneurial ecosystems should be tailored for different cohorts of nascent entrepreneurs as they require different support (Bell, 2019). Further integration of entrepreneurial ecosystems and EE can help reduce the leap and smoothen the transition from being entrepreneurial and starting a new venture.

References

Adler, P. S., & Kwon, S. W. (2002). Social capital: Prospects for a new concept. *The Academy of Management Review, 27*(1), 17–40.

Aldianto, L., Anggadwita, G., & Umbara, A. N. (2018). Entrepreneurship education program as value creation: Empirical findings of universities in Bandung, Indonesia. *Journal of Science and Technology Policy Management, 9*(3), 296–309.

Ally, M. (2004). Foundations of educational theory for online learning. In T. Anderson (Ed.), *The theory and practice of online learning* (pp. 16–44). Athabasca University Press.

Amer, A. (2006). Reflections on Bloom's revised taxonomy. *Electronic Journal of Research in Educational Psychology, 4*(1), 213–230.

Anderson, W. L., Krathwohl, D. R., Airasian, P. W., Cruikshank, K. A., Mayer, R. E., Pintrich, P. R., Raths, J., Wittrock, M. C. (2001). *A taxonomy for learning, teaching, and assessing: A revision of Bloom's taxonomy of educational objectives* (Complete Edition, 1st ed.). Pearson.

Aparicio, G., Iturralde, T., & Maseda, A. (2019). Conceptual structure and perspectives on entrepreneurship education research: A bibliometric review. *European Research on Management and Business Economics, 25*(3), 105–113.

Audretsch, D. B. (2014). From the entrepreneurial university to the university for the entrepreneurial society. *The Journal of Technology Transfer, 39*(3), 313–321.

Bandura, A. (1977). *Social learning theory.* Prentice-Hall.

Barr, S. H., Baker, T., Markham, S. K., & Kingon, A. I. (2009). Bridging the valley of death: Lessons learned from 14 years of commercialization of technology education. *Academy of Management Learning & Education, 8*(3), 370–388.

Béchard, J. P., & Toulouse, J. M. (1998). Validation of a didactic model for the analysis of training objectives in entrepreneurship. *Journal of Business Venturing, 13*(4), 317–332.

Bell, R. (2015). Developing the next generation of entrepreneurs: Giving students the opportunity to gain experience and thrive. *The International Journal of Management Education, 13*(1), 37–47.

Bell, R. (2016). Unpacking the link between entrepreneurialism and employability: An assessment of the relationship between entrepreneurial attitudes and likelihood of graduate employment in a professional field. *Education + Training, 58*(1), 2–17.

Bell, R. (2019), Predicting entrepreneurial intention across the university. *Education + Training, 61*(6/7), 815–831.

Bell, R. (2020a). Developing entrepreneurial behaviours in the Chinese classroom through value creation pedagogy. *Innovations in Education and Teaching International, 59*(1), 37–48.

Bell, R. (2020b). Adapting to constructivist approaches to entrepreneurship education in the Chinese classroom. *Studies in Higher Education, 45*(8), 1694–1710.

Bell, R. (2021). Underpinning the entrepreneurship educator's toolkit: Conceptualising the influence of educational philosophies and theory. *Entrepreneurship Education, 4*(1), 1–18.

Bell, R., & Bell, H. (2016a). An enterprise opportunity for entrepreneurial students: Student enterprise development and experience assessed through the student voice. *Education + Training, 58*(7/8), 751–765.

Bell, R., & Bell, H. (2016b). Replicating the networking, mentoring and venture creation benefits of entrepreneurship centres on a shoestring: A student-centred approach to entrepreneurship education and venture creation. *Industry and Higher Education, 30*(5), 334–343.

Bell, R., & Bell, H. (2020). Applying educational theory to develop a framework to support the delivery of experiential entrepreneurship education. *Journal of Small Business and Enterprise Development, 27*(6), 987–1004.

Bell, R. & Bell, H. (2023), Entrepreneurship education in the era of generative artificial intelligence. *Entrepreneurship Education.* https://doi.org/10.1007/s41959-023-00099-x

Bell, R., & Liu, P. (2019). Educator challenges in the development and delivery of constructivist active and experiential entrepreneurship classrooms in Chinese vocational higher education. *Journal of Small Business and Enterprise Development, 26*(2), 209–227.

Bennett, R. (2006). Business lecturers' perceptions of the nature of entrepreneurship. *International Journal of Entrepreneurial Behavior & Research, 12*(3), 165–188.

Biggs, J. (2012). What the student does: Teaching for enhanced learning. *Higher Education Research & Development, 31*(1), 39–55.

Blanka, C. (2019). An individual-level perspective on intrapreneurship: A review and ways forward. *Review of Managerial Science, 13*(5), 919–961.

Bloom, B. S. (1956). *Taxonomy of educational objectives: The classification of educational goals.* David McKay Company.

Blundel, R. (2002). Network evolution and the growth of artisanal firms: A tale of two regional cheese makers. *Entrepreneurship & Regional Development, 14*(1), 1–30.

Bridge, S. (2017). Is "entrepreneurship" the problem in entrepreneurship education? *Education + Training, 59*(7/8), 740–750.

Brieger, E., Arghode, V., & McLean, G. (2020). Connecting theory and practice: Reviewing six learning theories to inform online instruction. *European Journal of Training and Development, 44*(4/5), 321–339.

Cope, J. (2005). Toward a dynamic learning perspective of entrepreneurship. *Entrepreneurship Theory and Practice, 29*(4), 373–397.

Cranmer, S. (2006). Enhancing graduate employability: Best intentions and mixed outcomes. *Studies in Higher Education, 31*(2), 169–184.

Cui, J., Sun, J., & Bell, R. (2021). The impact of entrepreneurship education on the entrepreneurial mindset of college students in China: The mediating role of inspiration and the role of educational attributes. *The International Journal of Management Education, 19*(1), Article 100296.

Darkenwald, G. G., & Merriam, S. B. (1982). *Adult education: Foundations of practice*. Harper & Row.

Department for Business Enterprise and Regulatory Reform. (2008). *Enterprise: Unlocking the UK's talent*. HM Treasury.

Department for Education and Skills (DFES). (2003, January). *The future of higher education*. Paper presented to Parliament by the Secretary of State for Education and Skills.

Dierksmeier, C. (2020). From Jensen to Jensen: Mechanistic management education or humanistic management learning? *Journal of Business Ethics, 166*(1), 73–87.

Doutriaux, J., & Barker, M. (1996). University and industry in Canada: A changing relationship. *Industry and Higher Education, 10*(2), 88–102.

Education and Skills Committee. (2003, June). *The future of higher education*. The Stationery Office Limited.

Enterprise Educators UK (2021, August). *Timeline*. https://www.enterprise.ac.uk/about/#timeline

European Commission, Enterprise and Industry Directorate-General. (2008, March). *Entrepreneurship in higher education, especially within non-business studies*. European Commission. https://ec.europa.eu/docsroom/documents/8969/attachments/1/translations/en/renditions/pdf

Facione, P. (1990). *The California critical thinking skills test: College level*. California Academic Press.

Ferreira, C. C. (2020). Experiential learning theory and hybrid entrepreneurship: Factors influencing the transition to full-time entrepreneurship. *International Journal of Entrepreneurial Behavior & Research, 26*(8), 1845–1863.

Freire, P. (2006). Banking versus problem-solving models of education. In R. Curren (Ed.), *Philosophy of education: An anthology* (1st ed., pp. 68–75). Wiley-Blackwell.

Fuchs, K., Werner, A., & Wallau, F. (2008). Entrepreneurship education in Germany and Sweden: What role do different school systems play? *Journal of Small Business and Enterprise Development, 15*(2), 365–381.

Gedeon, S. (2014). Application of best practices in university entrepreneurship education: Designing a new MBA program. *European Journal of Training and Development, 38*(3), 231–253.

Gibb, A., & Hannon, P. (2006). Towards the entrepreneurial university. *International Journal of Entrepreneurship Education, 4*(1), 73–110.

Guerrero, M., & Peña-Legazkue, I. (2013). The effect of intrapreneurial experience on corporate venturing: Evidence from developed economies. *International Entrepreneurship and Management Journal, 9*(3), 397–416.

Halpern, D. F. (1996). *Thought and knowledge: An introduction to critical thinking*. Lawrence Erlbaum.

Hannon, P. D. (2005). Philosophies of enterprise and entrepreneurship education and challenges for higher education in the UK. *The International Journal of Entrepreneurship and Innovation, 6*(2), 105–114.

Hannon, P. D. (2006). Teaching pigeons to dance: Sense and meaning in entrepreneurship education. *Education + Training, 48*(5), 296–308.

Harima, A., Kroczak, A., & Repnik, M. (2021). Role ambiguity in entrepreneurship education: Expectation gaps between educators and students in venture creation courses. *Education + Training, 63*(9), 1309–1325.

Hassan, A., Saleem, I., Anwar, I. & Hussain, S.A. (2020). Entrepreneurial intention of Indian university students: the role of opportunity recognition and entrepreneurship education. *Education + Training, 62* (7/8), 843–861.

Henry, C. (2013). Entrepreneurship education in HE: Are policy makers expecting too much?. *Education + Training, 55*(8/9), 836–848.

Higher Education Funding Council Executive (2004), *HEFCE Strategic Plan for 2006–2011.*

Hindle, K. (2010). How community context affects entrepreneurial process: A diagnostic framework. *Entrepreneurship & Regional Development, 22*(7–8), 599–647.

Honig, B. (2004). Entrepreneurship education: Toward a model of contingency-based business planning. *Academy of Management Learning & Education, 3*(3), 258–273.

Hytti, U., & Kuopusjärvi, P. (2004). *Evaluating and measuring entrepreneurship and enterprise education: Methods, tools and practices.* Small Business Institute, Business Research and Development Centre, Turku School of Economics and Business Administration. https://ibw.at/resource/download/1419/entreva-net-report,pdf

Hytti, U., & O'Gorman, C. (2004). What is "enterprise education"? An analysis of the objectives and methods of enterprise education programmes in four European countries. *Education + Training, 46*(1), 11–23.

Ismail, A. B. T., Sawang, S., & Zolin, R. (2018). Entrepreneurship education pedagogy: Teacher-student-centred paradox. *Education + Training, 60*(2), 168–184.

Jack, S. L., & Anderson, A. R. (1999). Entrepreneurship education within the enterprise culture: Producing reflective practitioners. *International Journal of Entrepreneurial Behaviour & Research, 5*(3), 110–125.

Jonassen, D. H. (1991). Evaluating constructivistic learning. *Educational Technology, 31*(9), 28–33.

Jones, B., & Iredale, N. (2010). Enterprise education as pedagogy. *Education + Training, 52*(1), 7–19.

Jones, C. (2019). A signature pedagogy for entrepreneurship education. *Journal of Small Business and Enterprise Development, 26*(2), 243–254.

Jones, C., & English, J. (2004). A contemporary approach to entrepreneurship education. *Education + Training, 46*(8/9), 416–423.

Jones, S. (2014). Gendered discourses of entrepreneurship in UK higher education: The fictive entrepreneur and the fictive student. *International Small Business Journal, 32*(3), 237–258.

Kakouris, A., & Liargovas, P. (2021). On the about/for/through framework of entrepreneurship education: A critical analysis. *Entrepreneurship Education and Pedagogy, 4*(3), 396–421.

Kantar, L. D. (2013). Demystifying instructional innovation: The case of teaching with case studies. *Journal of the Scholarship of Teaching and Learning, 13*(2), 101–115.

Katz, J. A. (1991). The institution and infrastructure of entrepreneurship. *Entrepreneurship Theory and Practice, 15*(3), 85–102.

Kauppinen, A., Paloniemi, K., & Juho, A. (2019). Failed firm founders' grief coping during mentoring: Learning as the single catalyst of their restarting performance narratives. *The International Journal of Management Education*, Article 100288.

Kirschner, P. A., Sweller, J., & Clark, R. E. (2006). Why minimal guidance during instruction does not work: An analysis of the failure of constructivist, discovery, problem-based, experiential, and inquiry-based teaching. *Educational Psychologist, 41*(2), 75–86.

Klofsten, M., Fayolle, A., Guerrero, M., Mian, S., Urbano, D., & Wright, M. (2019). The entrepreneurial university as driver for economic growth and social change—Key strategic challenges. *Technological Forecasting and Social Change, 141*, 149–158.

Kolb, D. (1984). *Experiential learning.* Prentice Hall.

Krathwohl, D. R. (2002). A revision of Bloom's taxonomy: An overview. *Theory into Practice, 41*(4), 212–218.

Kreitzer, A., & Madaus, G. (1994). Empirical investigations of the hierarchical structure of the taxonomy. In L. Anderson & L. Sosniak (Eds.), *Bloom's taxonomy: A forty-year retrospective* (pp. 64–81). University of Chicago Press.

Kuratko, D. F. (2005). The emergence of entrepreneurship education: Development, trends, and challenges. *Entrepreneurship Theory & Practice, 29*(5), 577–597.

Kyrö, P. (2015). The conceptual contribution of education to research on entrepreneurship education. *Entrepreneurship & Regional Development, 27*(9–10), 599–618.

Lackéus, M. (2014). An emotion-based approach to assessing entrepreneurial education. *The International Journal of Management Education, 12*(3), 374–396.

Lackéus, M. (2018). "What is value?"—A framework for analyzing and facilitating entrepreneurial value creation. *Uniped, 41*(1), 10–28.

Lackéus, M., Lundqvist, M., & Middleton, K. W. (2016). Bridging the traditional-progressive education rift through entrepreneurship. *International Journal of Entrepreneurial Behavior & Research, 22*(6), 777–803.

Lackéus, M., & Middleton, K. W. (2015). Venture creation programs: Bridging entrepreneurship education and technology transfer. *Education + Training, 57*(1), 48–73.

Löbler, H. (2006). Learning entrepreneurship from a constructivist perspective. *Technology Analysis & Strategic Management, 18*(1), 19–38.

Lord, T., & Baviskar, S. (2007). Moving students from information recitation to information understanding: Exploiting Bloom's taxonomy in creating science questions. *Journal of College Science Teaching, 36*(5), 40–44.

Macht, S. A., & Ball, S. (2016). "Authentic alignment"—A new framework of entrepreneurship education. *Education + Training, 58*(9), 926–944.

Mathews, M. (2007). Constructivist pedagogy for the business communication classroom. *Journal of College Teaching & Learning, 4*(11), 99–106.

Matlay, H., & Carey, C. (2007). Entrepreneurship education in the UK: A longitudinal perspective. *Journal of Small Business and Enterprise Development, 14*(2), 252–263.

Mayhew, K., Deer, C., & Dua, M. (2004). The move to mass higher education in the UK: Many questions and some answers. *Oxford Review of Education, 30*(1), 65–82.

McKeown, J., Millman, C., Reddy Sursani, S., Smith, K., & Martin, L. M. (2006). Graduate entrepreneurship education in the United Kingdom. *Education + Training, 48*(8/9), 597–613.

Mcmullan, W. E., & Long, W. A. (1987). Entrepreneurship education in the nineties. *Journal of Business Venturing, 2*(3), 261–275.

Merriam, S. B. (1982). Some thoughts on the relationship between theory and practice. In S. B. Merriam (Ed.), *Linking philosophy and practice: New directions for continuing education* (pp. 87–91). Jossey-Bass.

Merriam, S. B., Caffarella, R. S., & Baumgartner, L. M. (2007). *Learning in adulthood: A comprehensive guide* (3rd ed.). Jossey-Bass.

Meyer, A. D., Aten, K., Krause, A. J., Metzger, M. L., & Holloway, S. S. (2011). Creating a university technology commercialisation programme: Confronting conflicts between learning, discovery and commercialisation goals. *International Journal of Entrepreneurship and Innovation Management, 13*(2), 179–198.

Miller, D. J., & Acs, Z. J. (2017). The campus as entrepreneurial ecosystem: The University of Chicago. *Small Business Economics, 49*(1), 75–95.

Moreland, N. (2006). *Entrepreneurship and higher education: An employability perspective.* The Higher Education Academy.

Morselli, D. (2018). Teaching a sense of initiative and entrepreneurship with constructive alignment in tertiary non-business contexts. *Education + Training, 60*(2), 122–138.

Mueller, S., & Anderson, A. R. (2014). Understanding the entrepreneurial learning process and its impact on students' personal development: A European perspective. *The International Journal of Management Education, 12*(3), 500–511.

Murphy, E. J. (2007). Prior learning assessment: A review of Bloom's taxonomy and Kolb's theory of experiential learning: Practical uses for prior learning assessment. *The Journal of Continuing Higher Education, 55*(3), 64–66.

Nabi, G., & Holden, R. (2008). Graduate entrepreneurship: Intentions, education and training. *Education + Training, 50*(7), 545–551.

National Committee of Inquiry into Higher Education. (1997). *High Education in the Learning Society: Report of the National Committee.* HMSO.

National Council for Graduate Entrepreneurship (NCGE). (2004). *Mapping of existing activity to support graduate entrepreneurs.* National Council for Graduate Entrepreneurship Report, Coventry, UK.

Neck, H., Brush, C., & Greene, P. (2014). *Teaching entrepreneurship: A practice-based approach.* Edward Elgar.

Newman, C., & Giles, C. (2005, December 10). 1976 Sterling crisis details made public. *Financial Times.* https://www.ft.com/content/93cb0260-68f4-11da-bd30-0000779e2340

O'Brien, E., M. Cooney, T., & Blenker, P. (2019). Expanding university entrepreneurial ecosystems to under-represented communities. *Journal of Entrepreneurship and Public Policy, 8*(3), 384–407.

Peltier, J. W., & Scovotti, C. (2010). Enhancing entrepreneurial marketing education: The student perspective. *Journal of Small Business and Enterprise Development, 17*(4), 514–536.

Pepin, M. (2012). Enterprise education: A Deweyan perspective. *Education + Training, 54*(8/9), 801–812.

Piercy, N. (2013). Evaluating experiential learning in the business context: Contributions to group-based and cross-functional working. *Innovations in Education and Teaching International, 50*(2), 202–213.

Pinchot, G. (1985). *Intrapreneuring: Why you don't have to leave the corporation to become an entrepreneur*. Harper & Row.

Pittaway, L., & Cope, J. (2007). Entrepreneurship education: A systematic review of the evidence. *International Small Business Journal, 25*(5), 479–510.

Prince, M. (2004). Does active learning work? A review of the research. *Journal of Engineering Education, 93*(3), 223–231.

Quality Assurance Agency for Higher Education. (2012). *Enterprise and entrepreneurship education: Guidance for UK higher education providers.* QAA.

Quality Assurance Agency for Higher Education. (2018, January). *Enterprise and entrepreneurship education: Guidance for UK higher education providers.* https://www.qaa.ac.uk/docs/qaas/enhancement-and-development/enterprise-and-entrpreneurship-education-2018.pdf?sfvrsn=15f1f981_8

Rae, D. (2005). Entrepreneurial learning: A narrative-based conceptual model. *Journal of Small Business and Enterprise Development, 12*(3), 323–335.

Ragins, B. R., Cotton, J. L., & Miller, J. S. (2000). Marginal mentoring: The effects of type of mentor, quality of relationship, and program design on work and career attitudes. *The Academy of Management Journal, 43*(6), 1177–1194.

Ramsgaard, M. B. (2018). Experiential learning philosophies of enterprise and entrepreneurship education. In D. Hyams-Ssekasi & E. F. Caldwell (Eds.), *Experiential learning for entrepreneurship* (pp. 3–18). Palgrave Macmillan.

Ramsgaard, M. B., & Christensen, M. E. (2018). Interplay of entrepreneurial learning forms: A case study of experiential learning settings. *Innovations in Education and Teaching International, 55*(1), 55–64.

Robertson, M., & Collins, A. (2003). Developing entrepreneurship in West Yorkshire: West Yorkshire universities' partnership and Business Start-Up@Leeds Met. *Education + Training, 45*(6), 303–307.

Robinson, S., Neergaard, H., Tanggaard, L., & Krueger, N. F. (2016). New horizons in entrepreneurship education: From teacher-led to student-centered learning. *Education + Training, 58*(7/8), 661–683.

Schön, D. (1983). *The reflective practitioner: How professionals think in action.* Basic Books.

Schramm, C. J. (2006). *The entrepreneurial imperative: How America's economic miracle will reshape the world.* Harper Business.

Schumpeter, J. (1942). *Capitalism Socialism and Democracy* (3rd ed.). Harper.

Seikkula-Leino, J., Ruskovaara, E., Ikavalko, M., Mattila, J., & Rytkola, T. (2010). Promoting entrepreneurship education: The role of the teacher?. *Education + Training, 52*(2), 117–127.

Sewell, P., & Pool, L. D. (2010). Moving from conceptual ambiguity to operational clarity: Employability, enterprise and entrepreneurship in higher education. *Education + Training, 52*(1), 89–94.

Sullivan, R. (2000). Entrepreneurial learning and mentoring. *International Journal of Entrepreneurial Behavior & Research, 6*(3), 160–175.

Urbano, D., & Guerrero, M. (2013). Entrepreneurial universities: Socioeconomic impacts of academic entrepreneurship in a European region. *Economic Development Quarterly, 27*(1), 40–55.

Vestergaard, L., Moberg, K., & Jørgensen, C. (2012). *Impact of Entrepreneurship Education in Denmark*. Young Enterprise Denmark.

Watkins, D., & Stone, G. (1999). Entrepreneurship education in UK HEIs: Origins, development and trends. *Industry and Higher Education, 13*(6), 382–389.

Chapter 9
Entrepreneurship Education in the US

Arthur K. Ellis, Leanna Aker, and Jeremy Delamarter

9.1 Introduction

Entrepreneurship education (EE) as a formal course of study is relatively new in American higher education. Among the earliest was a course in entrepreneurship introduced at Harvard University in 1947. However, schools and colleges of business and business administration in US universities can be traced back to the late 1800s. Students majoring in business could typically select from a range of options including accounting, marketing, sales, finance, management, etc. As a mark of growth and change, the 1980s saw the emergence of yet another domain: EE as a full-fledged major or course of study. Today, more than 200 colleges and universities in the USA offer such programs.

Nonetheless, these programs are novel enough to many people who might ask, "What is EE?" Descriptions and definitions abound. In a formal sense, EE is a course of study, both academic and professional, through which students are taught knowledge and skills designed to equip them with competencies for recognizing, creating, developing, and implementing business opportunities. Although it can be and is taught at primary and secondary levels, the focus here is on university-level teaching and learning, from undergraduate to graduate levels.

Many of the academic and theoretical ideas of EE derive from disciplines such as economics, psychology, philosophy, sociology, and anthropology. The applied areas primarily aim at business development through idea creation, innovative products, and new realms of services. In this sense, there is a parallel with other professional

A. K. Ellis (✉)
Center for Global Curriculum Studies, Seattle Pacific University, Washington, USA
e-mail: aellis@spu.edu

L. Aker
City University of Seattle, Washington, USA

J. Delamarter
Saint Martin's University, Washington, USA

X. Xu, *Comparative Entrepreneurship Education*,
https://doi.org/10.1007/978-981-99-1835-5_9

areas of study, including medicine, law, and education, each of which features "real world" connections as an integral part of effective preparation. Internships featuring "on the job" settings of different kinds are typically added as ways and means to apply classroom learning to real-world settings.

Although EE as formal coursework is fairly new, the idea of entrepreneurship has been around for a long time. We read about Cyrus McCormick (1809–1884), a Virginian who after several disappointing efforts, won a gold medal at the Crystal Palace Exhibition in London in 1851 for his invention of a mechanized grain reaper. Patents followed, demand emerged, and he founded the McCormick Harvesting Machine Company, which later became International Harvester. With little more training than a fourth-grade education, McCormick showed world-class entrepreneurial skills as both inventor and businessman. He perceived a "need" for mechanized farming to replace the many field hands it took to reap a harvest field of grain. Thus, he created, literally, a vehicle to meet the need. Much needs to be said about his resolve: he endured more failures than successes along the way. McCormick was a case study in tenacity.

Henry Ford's (1863–1947) formal education consisted of an eighth-grade diploma and a bookkeeping course he took at a commercial school. He had no formal training as a mechanical engineer, and it was said that he could not read a blueprint. As a young man living in a time when automobiles were beginning to appear, he saw opportunity. Automobiles were then owned by rich people, and Ford envisioned cars for ordinary folks. Cars were largely handmade in the early days, and Ford in fact made a couple. It seemed impossible to mass-produce cars at a reasonable cost; moreover, there were very few roads. Ford then went on to "invent" the assembly line, where individual workers specialized in limited aspects of production as cars moved along a belt, marking a huge step forward in efficiency. He also introduced the USD5 daily wage for assembly line workers, a rate unheard of at the time, thus ensuring the hiring of skilled people. The ripple effect of Ford's creativity was huge: it created the need for road construction, destination sites for leisure and industry, and many other examples of the multiple effects of disruptive change. The horse, for instance, was suddenly relegated from "indispensable" for farming and the military, to largely recreational pastimes duties.

These two legendary examples of entrepreneurship harken back to a time when brilliant people, often with little formal education and working alone, produced and implemented world-class ideas—McCormick in agriculture, and Ford in industry. The failure rate was high, and these two pioneers were exceptional. In the twenty-first century, global competitiveness has been a large factor in driving the idea that entrepreneurship can be taught, and that there are certain pathways that have proven to be fortuitous. There were two major changes: (1) the formal teaching of entrepreneurship at all levels of formal schooling, and (2) teamwork is more productive than individuals working alone.

Although the storied "lone heroes" model has a certain romantic appeal, it has essentially been replaced in modern times by team approaches. A recent example is the work by teams at Pfizer-BioNtech and Moderna for the rapid and successful development of vaccines for the Covid-19 virus. It was not a single name that took credit; rather, highly skilled individuals worked together as teams, sharing expertise, and coordinating efforts. While the idea of personal development remains, the power of social development is so overwhelmingly obvious that learning to work together is fundamental to any entrepreneurial efforts in the twenty-first century.

9.2 Teaching and Learning Entrepreneurship

If the underlying conditions of EE is setting a foundation of opportunity, needs assessment, creativity, out-of-the-box thinking, demand creation, initiative-taking, and action orientation, there is ample reason to think that the encouragement and development of academic, professional, and people skills must be part of the equation.

If we accept the idea that in the twenty-first century, random emergence of talented entrepreneurs is not a sufficient way to meet and create economic opportunity in an increasingly competitive world, then it follows that some form of systematic development is needed. The challenge for those who develop courses of study and who teach EE is how to organize a curriculum that combines knowledge, skills, and values that benefit both individuals and groups.

A curriculum can be "about," "in," or "through" certain knowledge, skills, and values. *How* a curriculum is centered is as important as the subject matter it purports to offer. For a typical EE course, there is the matter of what constitutes entrepreneurship, how best to go about being an entrepreneur, and what social/moral obligations are linked to entrepreneurship. The easy answer is that all three are important. But how teachers and students approach the knowledge, skills, and values, can and will vary greatly.

A curriculum or an entire program can be knowledge-centered, learner-centered, or society-centered. An astute observer can readily see how and to what extent these three options are featured, and/or to what extent they are neglected. Curricular balance, whether within a given course or developed throughout a program, is strategically important.

A knowledge-centric curriculum is familiar to almost any university student. The most common form of knowledge-centering at the university level is found in lectures and readings. The idea is that someone knows something that students do not know, or do not know well, and which they need to learn. It is a performer/audience model. It may be simplistic, but it is also temptingly realistic to say that telling is teaching. An expert knows certain required knowledge, and they tell students about it. Learning is assessed through examinations, typically written tests. Teachers vary greatly in their command of a subject as well as in their ability to make it interesting to learners. Students vary in their capacity and desire to learn through auditory means or via the written word. One of the arguments for "telling as teaching" is that it is efficient. It can be. Concepts are explained, material is covered, wisdom shared. But the words of Confucius should be heeded: "I hear, and I forget, I see and I remember, I do and I understand." To be fair, it is seldom the case that a class is based purely on listening to lectures and reading texts. Class discussion is sometimes held, but even then the flow of discussion is typically teacher-student, teacher-student.

A society-centered curriculum invites student-to-student interaction where they work in pairs, groups, etc. Class discussion is student-dominated, and students work together during class time as well as on assignments. The role of a student thus changes from receiver to participant. Project learning is emphasized, as are group skills. This often involves a "flipped classroom" model where readings/lectures are provided in advance, so that class time is devoted to discussion and other forms of active participation regarding the content or skills to be addressed. Assessment often takes the form of group performance or production. Teamwork is emphasized and an attempt is made to develop a class culture over time.

A third model—a society-centered curriculum—is one that makes every reasonable attempt to link content material with societal applications. An example is relocating a course, or a portion of it, to a real-world setting where the application of ideas is routine. Field trips, apprenticeships, projects that connect students with businesses are emphasized. In the field of teacher education, students will often say that they learned more in a semester of teaching than they did in their years of taking education classes. Whether they did or not is an open question, but the point is that application with guidance is an effective way to learn.

The point is not so much to denigrate or excessively praise any one of these models of curriculum. Rather, a program, in order to be effective, must be articulated in such a way that is balanced. In fact, each of the three models has its merits and limitations. But any program that purports to teach entrepreneurial knowledge, skills, and values must feature all three.

Can entrepreneurship be taught? Put another way, are entrepreneurs born or made? These are not idle or rhetorical questions. In fact, the debate of nature vs. nurture has had tangible effects on the systems, practices, and philosophies of EE programs around the world. Those on the extreme "nature" side argue that EE is only effective for students who have preexisting and innate entrepreneurial drives and dispositions (e.g., Radipere, 2012; Weber, 2013). If this is true, EE is not a matter of helping students learn new skills, but rather of fostering and freeing the tendencies that they already have. Thus, to say "she is an entrepreneur" is not a description of a person's behavior but rather a statement about a person's essence. An entrepreneur is an individual who exists in a state of entrepreneurial being, and entrepreneurship is an inherent quality, an a priori condition. Although an entrepreneurial disposition may be refined through education and experience, it cannot be introduced a posteriori. Indeed, there is some evidence to suggest that entrepreneurial dispositions are hardwired, at least to a degree (e.g., Carland & Carland, 2000; Dinis et al., 2013; Ferreira et al., 2012; Frese & Gielnik, 2014). Genetically determined lower levels of dopamine, for example, may compel an individual to seek out new and novel experiences such as starting a business, whereas a naturally heavily myelinated connection to the limbic system may lead to a fear-based aversion to risky behavior. Shane (2010) found that identical twins shared significantly more entrepreneurial traits than fraternal twins, suggesting that the shared genetic code among the identical twins was the source of their common entrepreneurial spirit.

The belief that entrepreneurial ability is genetically determined has profound implications for EE programs. Hayes and Richmond (2017), for example, highlight the use of quantitative entrepreneurial competency measures as tools for recruiting, marketing, and admissions in some American universities. These instruments are designed to function based on the assumption that entrepreneurship students are measurably different from other students, and, particularly, other business students *before* entering a program. Contrast these assumptions with emerging models in Nigeria that focus less on recruiting students with pre-identified abilities and more on building systems on the premise that "entrepreneurial learning should be conceived as a lifelong process, where knowledge is continuously shaped and revised as new experience take place" (Kulo et al., 2017, p. 50).

Attempts to invest an entire society with an entrepreneurial spirit and/or inclination through public school education is one thing. Identifying those with the greatest potential to benefit from advanced, university-level training is quite another. Almost anyone can be taught and can learn basic arithmetic skills. Therefore, mathematics is taught as a school subject at all primary and secondary schools. But not everyone has the interest, ability, or inclination to benefit from highly advanced courses in mathematics. This argument rests as a foundation for the very existence of higher education.

Specialization is a hallmark of advanced study, in particular. A field such as engineering demands knowledge and skills in mathematics and science that go well beyond basic coursework in calculus and physics. But beyond that lie specializations in mechanical, electrical, civil, and other branches of engineering, each of which requires highly specialized skillsets. Medicine requires special skills and mindsets beyond the MD level, and this is typically true of advanced specialized study in most spheres.

Lazear (2005), however, argued that entrepreneurs differ from specialists. Specialists have greater knowledge of a single skill, while entrepreneurs typically have a broader skillset. Specialists do well to work for others in high-capacity modes. In contrast, successful entrepreneurs are able to spot talent and combine a variety of skills. They have enough general knowledge to judge the qualities of applicants, for example. Successful entrepreneurs typically have a sense of the potential positive effects of the "whole," where "the whole is greater than the mere sum of the separate parts," to cite the gestalt theorem. The question for those who offer EE is how to select students with the most potential and those who have the greatest potential to enhance their "natural" proclivities.

Of course, arguments about nature vs. nurture in *any* educational setting are overly simplistic. Genetic expression is shaped and modified by environment, and genetically determined neural pathways can be altered, even radically, by experience. Consequently, the most recent and dominant discourse in entrepreneurial research acknowledges that entrepreneurial learning is a function of the interaction between nature and nurture (e.g., Arora, 2019). Contemporary studies, in fact, highlight the ability of EE to alter learners' psychological makeup, suggesting that although preexisting neural and psychological traits may indeed be linked to entrepreneurial ambitions, the impact of explicit and structured educational activities actually play a larger and more important role in fostering entrepreneurial behavior (e.g., Ndofirepi, 2020). In fact, even a positive perception of EE has been shown to have a positive correlation with learners' levels of innovation (Wei et al., 2019).

The question before us, then, is not whether entrepreneurship can be taught but rather how it can be taught in such a way that takes into account both learners' predispositions and their potential to develop. Effective EE must acknowledge individuals' inherent orientations while simultaneously providing structural and social influences that shape, develop, and reorient. In essence, entrepreneurial educators at the post-secondary level must give as much attention to learners' identities, relationships, and ways of being as they do to skill acquisition and knowledge development.

Of course, issues of identity and being are not unique to EE, but they are all the more important in a field where success is often considered to be the result of innate qualities as opposed to learned behaviors. What follows, then, is a review of constructivist andragogical frameworks used in EE that treat learning as a function of *being* and that do not draw hard distinctions between the development of skills and development of the self.

9.2.1 Transformative Learning

Mezirow (1991, 1997) posited that andragogy is less about the acquisition of skills or knowledge and more about learning to see the world in a new way. He referred to this change in orientation as "transformative learning," which is the process of effecting change in a *frame of reference.* Adults have acquired a coherent body of experience—associations, concepts, values, feelings, conditioned responses—that is, frames of reference that define their life world. Frames of reference are the structures of assumptions through which we understand our experiences. They selectively shape and delimit expectations, perceptions, cognition, and feelings. They set our "line of action." Once set, we automatically move from one specific activity (mental or behavioral) to another. We have a strong tendency to reject ideas that fail to fit our preconceptions, labeling those ideas as unworthy of consideration—aberrations, nonsense, irrelevant, weird, or mistaken. When circumstances permit, transformative learners move toward a frame of reference that is more inclusive, discriminating, self-reflective, and integrative of experience (Mezirow, 1997, p. 5).

The process of transformative learning begins with an external catalyst, sometimes referred to as a "disorienting dilemma" or a "critical encounter." This confrontation with the unexpected causes the learner to question not just her knowledge and skills but also her values and her unquestioned assumptions about the nature of the world. She enters into a period of self-examination and alienation. This inward-focused critique eventually leads to exploration of new behavior, the development of new skills and knowledge, experimentation with new roles, and finally a reconceptualization of the self. This reconceptualization is accompanied by new practices and orientations. It is both practically and metaphysically a new way of being.

Entrepreneurship as a potentially transformative space is a relatively new concept. As recently as 2011, Kakouris has highlighted the field's general lack of pedagogy to address students' personal assumptions. He claims, however, that entrepreneurship is a field rife with myths and assumptions:

> [There are] a series of 'common sense' beliefs that contradict with worldwide relevant data on business venturing. As a consequence, educators have to confront beliefs (or 'myths') in class in order to foster an entrepreneurial mindset to trainee populations of diverse entrepreneurial biases. (Kakouris, 2011, p. 653)

Despite the need for entrepreneurship programs to actively help students confront and revise their unquestioned presuppositions, such opportunities have been "poorly exploited" (Kakouris, 2011, p. 653).

Fortunately, the last decade has seen an explosion in research relating to transformative learning in EE, which highlights both its growing popularity as a model and its efficacy in fostering sustainable entrepreneurial outcomes. Both scholars and educators have becoming increasingly aware of the need for a transformative approach to EE, and the data have revealed two key transformative andragogical practices.

First, EE programs must intentionally and programmatically expose their students to disorienting dilemmas, thereby kickstarting the transformation process (e.g., Aboytes et al., 2022; Ndlela et al., 2019; Nyamunda & Van Der Westhuizen, 2020). Neergaard et al. (2020) refer to this process as "pedagogical nudging," which leads students to interrogate their sense of belonging in the entrepreneurial field. Although it is possible that students will naturally encounter disorienting dilemmas during internships or other field-based experiences, courses or programs are actually more effective for triggering productive reflection and growth (Timmer, 2015), and it is imperative for programs to introduce them in safe and controllable ways. It is important to note that these introductions need not be complicated. For example, Benson et al. (2012) demonstrated the effectiveness of a simple homework assignment where students had to dress in clothes they would not normally wear. This experience, and the reactions it provoked from friends and family, was enough to serve as a disorienting dilemma for many students in the class. The professors followed up with journaling and discussion exercises where students reported "feelings of confusion, guilt, liberation, or even gaining new insights" (Benson et al., 2012, p. 155). These reflections led to new perspectives on customer engagement and were ultimately incorporated into students' entrepreneurial plans.

As demonstrated above, the follow-up activity is as important as the disorienting dilemma itself, and it is the second trend identified in existing literature: programs must provide a sustained and structured space for reflection following the disorienting dilemma (e.g., George, 2015; Hytti et al., 2020; Vettraino & Jones, 2022). Classroom-based reflective activities centred on entrepreneurial experiences have been demonstrated to positively impact learners' self-efficacy (Kassean et al., 2015), perception of skills and abilities (Longva, et al., 2020), and meaning-making (Bell & Bell, 2020). Reflection is most effective when paired with experience and opportunities for action (Hägg & Kurczewska, 2016), particularly when conducted in a team setting in response to a disorienting dilemma (Scott et al., 2020).

9.2.2 Communities of Practice

Of course, transformative learning typically takes place within a social context. Learning, Lave and Wenger (1991) argue, is always socially situated. The idea of *becoming*, then, is less a matter of (re)aligning one's individual values based on new experience and information, but more a matter of being drawn into group membership through shared practice. This process of being "drawn in" requires intentional interactions between newcomers and the community's established members:

> Practice is a shared history of learning that requires some catching up for joining. It is not an object to be handed down from one generation to the next. Practice is an ongoing, social, interactional process, and the introduction of newcomers is merely a version of what practice already is. That members interact, do things together, negotiate new meanings, and learn from each other is already inherent in practice—that is how practices evolve… [and] communities of practice reproduce their membership in the same way that they come about in the first place. (Wenger, 1998, p. 33)

Newcomers begin on the periphery of the community, observing and taking part in "legitimate peripheral practices" (Lave & Wenger, 1991). Over time and through sustained interaction with the community, they learn to do as the community does. They adopt the habits common to members of the community—the dress, vocabulary, references, and even epistemological approaches. As a result, they learn to *be* by first learning to *do*.

Situated learning theories and communities of practice form the foundation of many EE activities including networking (Lefebvre et al., 2015), business coaching (Haneberg & Aaboen, 2021), customer relations (Terzieva, 2016), employee development (Alzoubi, 2021), brand communication (Holikatti et al., 2019), and fundraising (Gautier et al., 2021).

The deliberate inclusion of situated learning theories into EE allows program leaders and instructors to create meaningful, practice-based learning opportunities for students that compel them to interact with members of the broader entrepreneurial community. These interactions, often in the form of internships and similar mentorship models, benefit not only the student and the mentor but the educational program as well (Gautier et al., 2021). Recent research highlights the role of communities of practice in helping EE programs shift to online and blended learning models (e.g., Hafeez et al., 2018; Kineshanko & Jugdev, 2018; Polbitsyn et al., 2020), use AI-assisted pedagogical tools (e.g., Ala et al., 2022), develop and teach cybersecurity measures (e.g., Nobles & Burrell, 2018), and develop ethical frameworks (e.g., Weller, 2020).

9.2.3 Constructivist Learning Models

A community of practice model of EE is predicated upon an action-oriented pedagogy in which learning is not transactional but constructed. The constructivist foundation of hands-on, experiential learning is well documented (e.g., Bruner, 1960; Dewey, 1938), and it is beyond the scope of this chapter to recount the basics of constructivist theories. It is important, however, to note that the pedagogical trends highlighted in research on effective EE are prototypically constructivist. Both the transformative learning theory and communities of practice foreground learners' active engagement in the learning process. Both highlight the socio-relational nature of learning, and the need for a physical and reflective space where the learner can practice emerging ways of being.

However, although contemporary research is more or less unanimous about the efficacy of and need for constructivist approaches to EE, contemporary practice has been slow to change. For example, multiple studies have highlighted the challenges faced by Chinese EE programs as they attempt to move away from passive learning models and toward constructivist, enactive models (e.g., Bell, 2020; Bell & Liu, 2018). Similarly, recent studies of European entrepreneurship programs revealed that the programs were not guided by any shared, agreed-upon constructivist principles (Baggen et al., 2021). The development and wide adoption of constructivist approaches to EE is perhaps the single most effective step that entrepreneurial programs can take toward preparing students for contemporary entrepreneurial opportunities (e.g., Bell, 2022; Curtis et al., 2021; Löbler et al., 2021; Rasiah et al., 2019; Wasim, 2019), and constructivist approaches like those described above are uniquely situated to foster both the development of skills and of the self.

9.3 Exemplary EE Programs at US Universities

9.3.1 Description of the EE Curriculum at Babson University

Babson College (main campus in Wellesley, Massachusetts, with branches in Boston, MA, San Francisco, CA and Miami, FL) is a private post-secondary educational institution focused on business in general, and entrepreneurship in particular. Babson College offers degrees at the undergraduate level (Bachelor of Science), at the graduate level (Master of Business Administration and Master of Science), and certificates at the executive education level. The Babson College graduate programs include a Master of Business Administration (MBA), Master of Science (MS) in Management in Entrepreneurial Leadership, MS in Finance, MS in Business Analytics, and Certificate in Advanced Management, Executive Entrepreneurial Leadership Certificate, and Executive Certificate in Global Business Leadership.

The institutional mission statement illustrates a focus on entrepreneurship: "Babson College is the educator, convener, and thought leader for Entrepreneurship of All Kinds®." US News and World Report currently ranks Babson College as USA's top college for entrepreneurship at the MBA and undergraduate levels. Babson has held the top ranking in both categories for almost 30 years.

The Babson College undergraduate program is a four-year business degree with an emphasis on the theory and practice of entrepreneurial approaches to business. EE in the Babson context is differentiated from the traditional elements of business school training such as management, marketing, finance, and accounting. While students receive instruction in these more traditional business school areas, Babson keeps curricular focus on entrepreneurship theory and practice.

Babson's curricular philosophy emphasizes two main elements: interdisciplinary teaching and project-based learning. Interdisciplinary teaching at Babson utilizes faculty from different business disciplines (marketing, finance, accounting, etc.) who use case studies to help students analyze and understand how business theories are operationalized in real-world contexts. Project-based learning allows students to learn by doing, offering teams of students the opportunity to take on real-world entrepreneurial tasks as part of their coursework. The strongest example of project-based learning, and a feature that Babson is known for, is the Foundations of Management and Entrepreneurship course.

The following is a description of the Entrepreneurship Concentration at Babson University:

> The Entrepreneurship Concentration focuses on the creation of social and economic value by developing core capacities of idea generation, opportunity recognition, resource acquisition, and entrepreneurial management. Entering students will learn to shape entrepreneurial opportunities, assess financial feasibility, while living an entrepreneurial experience.

The year-long Foundations of Management and Entrepreneurship course is a requirement for all first-year students. Students work in teams on a project to identify and create a novel product or service that has potential for profitability. Teams are loaned up to $3,000 in startup funds, and professors guide students through all phases of taking a business startup from idea to completion. Local nonprofit agencies benefit from these projects by receiving volunteer hours from the students as well as sharing 50% of any project-generated profit. This course allows students to contextualize and apply the theoretical aspects of entrepreneurial management, and builds a base for the subsequent three years of undergraduate coursework.

In addition to this project-based first-year experience, the Babson curriculum emphasizes authentic assessment through student projects and products as well as industry peer evaluators. These assessments are aligned with the type of tasks one might encounter in a business setting as a nascent entrepreneur. For example, students might be expected to produce and present business pitches or elevator speeches, design product prototypes, research and write consulting reports for local enterprises, and/or attend and present at industry trade shows. These projects or products are often evaluated by a recruited panel of business owners or industry representatives who provide assessment and feedback. This feedback provided by industry representatives gives students real-world assessment and opportunities for networking.

9.3.2 UCLA Entrepreneurship Programs

The University of California Los Angeles (UCLA) offers EE through an undergraduate minor, a concentration in the Master of Business administration (MBA) degree, certificate programs, competitions, collaborations, fellowships, and internships. The MBA degree offers options for students to attend school full time, or to attend part-time while remaining fully employed (Anderson School of Management, n.d.-a). The length of both programs is roughly two years, ranging from 24 to 28 credits. The entrepreneurship minor is available to UCLA students across colleges and degree programs, with a total of 20 credits (UCLA, n.d.-a).

UCLA offers less traditional EE through fellowships, competitions, and networking opportunities. The Harold and Pauline Price Center for Entrepreneurship and Innovation (PCEI) is the epicenter of entrepreneurship at UCLA. It claims to "depart from traditional business studies by emphasizing social innovation and producing managers who know how to marshal resources for new wealth creations" (Anderson School of Management, n.d.-b). For example, the Venture Accelerator is a six-month immersive program for early-stage startups and their founders (Anderson School of Management, n.d.-c). PCEI also has an entrepreneur association of over 600 members which hosts networking events and workshops (Anderson School of Management, n.d.-b).

While an exploration of program offerings gives a broad sense of EE at UCLA, a finer-grained exploration of courses affords a more nuanced view. Courses are offered in traditional academic formats as well as field-based capstones and internships. Course titles include Technology for Competitive Advantage, Entrepreneurship and Business Plan Development, Persuasion and Influence, and Social Entrepreneurship (Anderson School of Management, n.d.-d). The MBA degree with an entrepreneurship concentration includes two courses of fieldwork, and provides an option to for students to participate in Business Creation Option (BCO). BCO supports students in launching their own businesses (Anderson School of Management, n.d.-e). The capstone course for the undergraduate minor in entrepreneurship includes fieldwork and internship components.

The curriculum in the school of management at UCLA emphasizes values-based decision making as well as global experience and knowledge. Edwards, a chair in the center for global management, states that "there's no way that you can be a successful business person in the twenty-first century if you don't know the world" (Anderson School of Management, n.d.-f). MBA students must fulfill a global option through one of three tracks: (1) participation in a global immersion course (virtual or with travel), (2) completion of international research or BCO option, or (3) enrollment in global management electives (Anderson School of Management, n.d.-f). Additionally, students take immersive courses in sustainability, equity, and social impact (Anderson School of Management, n.d.-g).

Entrepreneurship courses at UCLA seem to employ traditional higher education assessments, such as papers, exams, projects, and participation. For example, a sample syllabus from the entrepreneurship and new product development course shows the following assessments and grading breakdown: exams (30%), cases (25%), final project presentation (25%), class participation (15%), and team participation (5%) (Rostamian, 2019). Another course, Entrepreneurship and New Venture Formation, includes the following assessments: attendance, research assignment, written business plan, and business plan presentation. Scoring criteria for the business are holistic in nature. For example, a student would earn a score of a B+ if they had a "very good plan, adequately analyzed, but conclusions not fully supported" (UCLA Extension, 2018). Presumably, students' participation in some of the less traditional portions of the entrepreneurship offerings at UCLA (such as BCO and venture competitions) would reflect a different kind of performance assessment.

9.3.3 Harvard University Business School Entrepreneurship Programs

Harvard Business School (HBS) offers EE through degree and certificate programs, workshops, competitions, and mentoring/networking structures. The MBA degree includes an entrepreneurship specialization option with one year of required coursework under the guidance of an entrepreneurship manager, and a second year of elective coursework (Harvard Business School, n.d.-a). HBS also offers a certificate program in entrepreneurship and innovation, comprising four online courses ranging from four to eight weeks.

HBS offers less traditional EE through fellowships, competitions, mentoring, networking, coaching, pro bono legal advice, and financial support. For example, Rock Fellows is a 12-week intense summer program for second-year MBA students to work on venture ideas or nascent startups (Harvard Business School, n.d.-b). The program involves cohort meetings and individual time to develop tactics, meet customers, and engage in coaching with experts. Competitions for seed capital are part of similar workshops at HBS, and such financial support is a unique offering in EE (Harvard Business School, n.d.-c). Informal mentoring opportunities are provided through the entrepreneurs-in-residence program, which involves successful entrepreneurs offering one-on-one coaching and program support roughly six times a year (Harvard Business School, n.d.-d).

Courses are conducted in both a traditional academic format as well as in experiential field settings. Course titles include Founders' Journey, Disruptive Strategy, Launching Technology Ventures, and Avoiding Startup Failure (Harvard Business School, n.d.-e). Some course titles appear to reflect overlapping or similar topics found in more traditional MBA programs. Professor Howard Stevenson, considered the "godfather of entrepreneurship studies" at HBS, defines entrepreneurship as "the pursuit of opportunity beyond resources controlled" (Stevenson & Gumpert, 1985). Thus, even potentially overlapping courses such as Data for Impact can have an entrepreneurial slant.

Many courses in HBS employ a case method approach, and the entrepreneurship specialization courses are no different. HBS describes this approach as spontaneous and unscripted with the instructor serving as "planner, host, moderator, devil's advocate, fellow-student, and judge" (Christensen Center for Teaching and Learning, n.d.). Students and instructors co-create the course content and processes using this approach. Typically, students and instructors explore a particular case, with participation in discussions being a key component. In the Disruptive Strategies course, for example, the case method is applied through a team discussion of the blockbuster's demise and an exploration of the Godrej Group, which innovated around refrigeration issues in India (Harvard Business School, n.d.-f).

Most of the entrepreneurship courses at HBS employ traditional higher education assessment such as papers, exams, projects, and participation. However, assessment may be more authentic than what is gathered from a review of the course descriptions in the course catalog. For example, in the field course, Scaling Minority Businesses, the project involves working with a black-owned business to engage with a challenge or growth opportunity that the business is experiencing (Harvard Business School, n.d.-g). The course description states that assessment will be based on: "(1) quality of team's final presentation and viability of recommendations, (2) class participation, (3) peer evaluation, (4) situation analysis, (5) draft presentation, and (6) personal reflection" (Harvard Business School, n.d.-g). It is unclear if assessment criteria for these categories will be co-created or provided through a pre-determined rubric.

Authentic assessment can be seen through some of the more innovative program structures such as venture competitions, seed capital opportunities, and entrepreneurs-in-residence mentorship. Students receive authentic assessment if they obtain seed capital or fail to win a venture competition, for example. As each of these program structures includes mentorship and faculty guidance, students receive valuable, authentic feedback as part of their participation.

9.3.4 Best Programs

It could be argued that any determination of the "best" EE program is a subjective matter, and that it is the one that worked best for a given individual. It is estimated that currently more than 5,000 courses and hundreds of programs are available. And given the steep rise in tuition rates in recent years, for many students, affordability is always something to consider. Rising student debt is beyond the scope of this chapter, but any discussion of program quality has to take it into account. Few of the programs are inexpensive. Support of different kinds is often available through scholarships, internships, and, of course, student loans. Suitability to individual needs and interests is an important factor, and the best advice for anyone considering this area of study is to examine different programs carefully in order to make the choice that would most likely enhance the considerable investment of time and money spent in hopes of launching a rewarding career.

Whatever the case, higher education is a proven investment in a better future—financially, socially, and intellectually. The list of highly prestigious programs that follows is a mere sampling of the best of many very good EE programs.

US News & World Report (USNWR) has published lists of the quality of education in public and private universities in America. Any attempt to rank university programs is controversial, but given the systematic and thorough range of variables taken into account by USNWR, it is hard to ignore their findings. The following is the USNWR 2022 ranking of the top EE programs in universities in the US.

1. Babson University
2. Massachusetts Institute of Technology
3. University of California, Berkeley
4. Indiana University
5. University of Michigan
6. University of Pennsylvania
7. University of Texas, Austin
8. University of Utah
9. Baylor University
9. University of North Carolina, Chapel Hill
9. Santa Clara University

Other schools that finish regularly in the top tier are Stanford University, Harvard University, and Yale University. The schools are a mix of public and private universities. At least 200 universities in the United States, and even private training firms offer EE programs.

Criteria taken into account in the rankings include expert opinion in the form of peer assessment surveys, proportion of fulltime faculty, graduation and retention rates, class sizes, resources, student selection, career placements, spending per fulltime student, and alumni giving rates.

9.4 Conclusion

This chapter has defined and described EE in higher education, identified its key concepts and processes, and provided exemplary programs. It is clear that in recent years universities have made effective efforts to close the gap between the theoretical worlds of economics, business, psychology, and other academic/professional fields, and the "real world" of enterprise and commerce. Internships and adjunct professorships are but two examples of this. An internship allows joint guidance by university personnel and business owners as students apply ideas and strategies in the marketplace. Additionally, inviting successful entrepreneurs to guest lecture and/or teach adjunct courses brings the real world into the university.

The social/moral aspect of EE beyond the classroom is seen as a crucial aspect of growth and development as a result of multiple avenues of engagement, in the form of student clubs, competitions, alumni centers, institutes, campus infrastructure, experiential learning, authentic assessment techniques, and university-wide interdisciplinary programs.

Concepts such as value creation, innovation, market orientation, teambuilding, continuous growth, business culture, decision making, and clear vision take on operational definitions through application and trial and error, but they are grounded in thought and reflection through years of study. EE is based on the premise that preparation for an applied field involves serious study and academic learning, as well as social and moral growth designed to benefit students and the people they serve in the course of a career.

EE shares with teaching, medicine, and law an application component. It is something one learns, but also something that one continually learns by doing it. As Aristotle wrote, "For the things we have to learn before we can do them, we need to learn by doing them."

References

Aboytes, J. G. R., Barth, M., & Fischer, D. (2022). Evolution of entrepreneurs' expectations using instagram as a business practice: A transformative learning perspective in the case of sustainable fashion entrepreneurs in Mexico. *World Development Sustainability, 1*, Article 100005.

Ala, M., Robin, M., Rasul, T., & Wegner, D. (2022). Understanding the possibilities and conditions for instructor-AI collaboration in entrepreneurship education. *Technology and entrepreneurship education: Adopting creative digital approaches to learning and teaching* (pp. 159–185). Palgrave Macmillan.

Alzoubi, M. O. (2021). Exploring the relationships among tacit knowledge sharing, communities of practice and employees' abilities: The case of KADDB in Jordan. *International Journal of Organizational Analysis*. https://doi.org/10.1108/IJOA-11-2020-2480

Anderson School of Management, UCLA. (n.d.-a). *UCLA Anderson degree programs*. Retrieved May 20, 2022 from https://www.anderson.ucla.edu/degrees

Anderson School of Management, UCLA. (n.d.-b). *Price Center for Entrepreneurship and Innovation: For students*. Retrieved May 20, 2022 from https://www.anderson.ucla.edu/about/centers/price-center-for-entrepreneurship-and-innovation/for-students

Anderson School of Management, UCLA. (n.d.-c). *Anderson venture accelerator*. Retrieved May 20, 2022 from https://www.anderson.ucla.edu/about/centers/price-center-for-entrepreneurship-and-innovation/anderson-venture-accelerator).

Anderson School of Management, UCLA. (n.d.-d). *Undergraduate minor in entrepreneurship*. Retrieved May 20, 2022 from https://www.anderson.ucla.edu/about/centers/price-center-for-entrepreneurship-and-innovation/for-students/undergraduate-minor-in-entrepreneurship

Anderson School of Management, UCLA. (n.d.-e). *Business creation option*. Retrieved May 20, 2022 from https://www.anderson.ucla.edu/degrees/full-time-mba/academics/capstone-project/business-creation-option

Anderson School of Management, UCLA. (n.d.-f). *Going global: Becoming attuned*. Retrieved May 20, 2022 from https://www.anderson.ucla.edu/degrees/full-time-mba/academics/global-options

Anderson School of Management, UCLA. (n.d.-g). *Courses*. Retrieved May 20, 2022 from https://www.anderson.ucla.edu/about/centers/impactanderson/for-students/courses

Arora, K. (2019, March 21). Are successful entrepreneurs born or made? *Forbes*. https://www.forbes.com/sites/theyec/2019/03/21/are-successful-entrepreneurs-born-or-made/?sh=6696a0b03ed1

Baggen, Y., Lans, T., & Gulikers, J. (2021). Making entrepreneurship education available to all: Design principles for educational programs stimulating an entrepreneurial mindset. *Entrepreneurship Education and Pedagogy, 5*(3), 347–374.

Bell, R. (2020). Adapting to constructivist approaches to entrepreneurship education in the Chinese classroom. *Studies in Higher Education, 45*(8), 1694–1710.

Bell, R. (2022). Supporting students and society: Underpinning entrepreneurship education with a humanistic philosophy. In G. J. Larios-Hernandez, A. Walmsley, & I. Lopez-Castro (Eds.), *Theorising undergraduate entrepreneurship education* (pp. 157–172). Palgrave Macmillan.

Bell, R., & Bell, H. (2020). Applying educational theory to develop a framework to support the delivery of experiential entrepreneurship education. *Journal of Small Business and Enterprise Development, 27*(6), 987–1004.

Bell, R., & Liu, P. (2018). Educator challenges in the development and delivery of constructivist active and experiential entrepreneurship classrooms in Chinese vocational higher education. *Journal of Small Business and Enterprise Development, 26*(2), 209–227.

Benson, C. C., Palin, G. R., Cooney, T. M., & Farrell, K. (2012). Agents of change: Using transformative learning theory to enhance courses in business ethics & social entrepreneurship. *Journal of Legal Studies in Business, 22*, 97–167.

Bruner, J. (1960). *The process of education*. Harvard University Press.

Carland, J. C., & Carland, J. W. (2000). A new venture creation model. *Journal of Business and Entrepreneurship, 12*(3), 29–49.

Christensen Center for Teaching and Learning. (n.d.). *Teaching by the case method: Case. Method in practice*. Retrieved May 9, 2022 from https://www.hbs.edu/teaching/case-method/Pages/default.aspx

Christie, M., Carey, M., Robertson, A., & Grainger, P. (2015). Putting transformative learning theory into practice. *Australian Journal of Adult Learning, 55*(1), 9–30.

Curtis, V., Moon, R., & Penaluna, A. (2021). Active entrepreneurship education and the impact on approaches to learning: Mixed methods evidence from a six-year study into one entrepreneurship educator's classroom. *Industry and Higher Education, 35*(4), 443–453.

Dewey, J. (1938). *Experience and education*. Kappa Delta Pi.

Dinis, A., Paço, A., Ferreira, J., Raposo, M., & Rodrigues, R. G. (2013). Psychological characteristics and entrepreneurial intentions among secondary students. *Education + Training, 55*(8/9), 763–780. https://doi.org/10.1108/ET-06-2013-0085

Ferreira, J. J., Raposo, M. L., Rodrigues, R. G., Dinis, A., & Paço, A. D. (2012). A model of entrepreneurial intention: An application of the psychological and behavioural approaches. *Journal of Small Business and Enterprise Development, 19*(3), 424–440. https://doi.org/10.1108/14626001211250144

Frese, M., & Gielnik, M. M. (2014). The psychology of entrepreneurship. *Annual Review of Organizational Psychology and Organizational Behavior, 1*, 413–438. https://doi.org/10.1146/annurev-orgpsych-031413-091326

Gautier, A., Pache, A. C., Chowdhury, I., & Ligonie, M. (2021). The missing link: Communities of practice as bridges between institutional entrepreneurs and frontline practitioners in institutionalizing a divergent practice. In M. Lounsbury, D. Anderson, & P. Spee (Eds.), *On practice and institution: New empirical directions* (pp. 199–230). Emerald Publishing Limited.

George, B. (2015). Using transformative learning theory to develop a course in sustainable entrepreneurship. In V. L. Crittenden, K. Esper, R. Slegers, & N. Karst (Eds.), *Evolving entrepreneurial education: Innovation in the Babson classroom* (pp. 283–303). Emerald Publishing Limited.

Hafeez, K., Foroudi, P., Nguyen, B., Gupta, S., & Alghatas, F. (2018). How do entrepreneurs learn and engage in an online community-of-practice? A case study approach. *Behaviour & Information Technology, 37*(7), 714–735.

Hägg, G., & Kurczewska, A. (2016). Connecting the dots–A discussion on key concepts in contemporary entrepreneurship education. *Education + Training, 58*(7/8), 1–16. https://doi.org/10.1108/ET-12-2015-0115

Haneberg, D. H., & Aaboen, L. (2021). Entrepreneurial learning behaviour of community insiders. *International Journal of Entrepreneurial Behavior & Research, 28*(2), 306–324.

Harvard Business School. (n.d.-a). *Entrepreneurship courses*. Retrieved May 9, 2022 from https://entrepreneurship.hbs.edu/programs/mba/Pages/courses.aspx

Harvard Business School. (n.d.-b). *Entrepreneurship: Rock summer fellows*. Retrieved May 9, 2022 from https://entrepreneurship.hbs.edu/programs/mba/Pages/rock-summer-fellows.aspx

Harvard Business School. (n.d.-c). *Entrepreneurship: Rock accelerator*. Retrieved May 9, 2022 from https://entrepreneurship.hbs.edu/programs/mba/Pages/rock-accelerator.aspx

Harvard Business School. (n.d.-d). *Entrepreneurship: Entrepreneurs-in-residence.* Retrieved May 9, 2022 from https://entrepreneurship.hbs.edu/programs/mba/Pages/entrepreneurs-in-residence.aspx

Harvard Business School. (n.d.-e). *Elective curriculum: Course descriptions.* Retrieved May 9, 2022 from https://www.hbs.edu/coursecatalog/

Harvard Business School. (n.d.-f). *Harness the power of disruptive innovation to craft winning strategies.* Retrieved May 9, 2022 from https://online.hbs.edu/courses/disruptive-strategy/?utm_source=harvard-online&utm_medium=affiliate#syllabus

Harvard Business School. (n.d.-g). *HBS course catalog: Field course: Scaling minority businesses.* Retrieved May 9, 2022 from https://www.hbs.edu/coursecatalog/6614.html

Hayes, D., & Richmond, W. (2017). Using an online assessment to examine entrepreneurship student traits and to measure and improve the impact of entrepreneurship education. *Journal of Entrepreneurship Education, 20*(1), 88–107.

Holikatti, M., Jhaver, S., & Kumar, N. (2019). Learning to Airbnb by engaging in online communities of practice. *Proceedings of the ACM on Human-Computer Interaction, 3*(CSCW), 1–19.

Hytti, U., Heinonen, J., & Stenholm, P. (2020). Entrepreneurship for research professionals: Triggering transformative learning? In L. Aaboen, H. Landström, & R. Sørheim (Eds.), *How to become an entrepreneur in a week* (pp. 100–118). Edward Elgar Publishing.

Kakouris, A. (2011). Innovative entrepreneurship: A new field that urges for transformative learning? In M. Alhadeff-Jones & A. Kokkos (Eds.), *Transformative learning in time of crisis: Individual and collective challenges.* Proceedings of the 9th International Transformative Learning Conference, New York, NY & Athens, Greece. https://www.academia.edu/787737/Transformative_Learning_in_Time_of_Crisis_Individual_and_Collective_Challenges

Kassean, H., Vanevenhoven, J., Liguori, E., & Winkel, D. E. (2015). Entrepreneurship education: A need for reflection, real-world experience and action. *International Journal of Entrepreneurial Behavior & Research, 21*(5), 690–708.

Kineshanko, M. K., & Jugdev, K. (2018). Enhancing digital intelligence through communities of learning. In A. Khare & D. Hurst (Eds.), *On the line: Business education in the digital age* (pp. 111–125). Springer.

Kulo, V. A., Agbogo, R. A., & Okudarie, J. U. (2017). Challenges of entrepreneurship education in Nigeria. *Nigerian Journal of Business Education (NIGJBED), 4*(1), 50–56.

Lave, J., & Wenger, E. (1991). *Situated learning: Legitimate peripheral participation.* Cambridge University Press. https://doi.org/10.1525/ae.1994.21.4.02a00340

Lazear, E. P. (2005). Entrepreneurship. *Journal of Labor Economics, 23*(4), 649–680.

Lefebvre, V., Radu Lefebvre, M., & Simon, E. (2015). Formal entrepreneurial networks as communities of practice: A longitudinal case study. *Entrepreneurship & Regional Development, 27*(7–8), 500–525.

Löbler, H., Maier, M., & Markgraf, D. (2021). Evaluating the constructionist approach in entrepreneurship education. *Research in Teacher Education, 11*(1), 83–89.

Longva, K. K., Strand, Ø., & Pasquine, M. (2020). Entrepreneurship education as an arena for career reflection: The shift of students' career preferences after a business planning course. *Education+ Training, 62*(7/8), 877–896.

Man, T. W. Y. (2019). Nurturing entrepreneurial competencies through university-based entrepreneurship centers: A social constructivist perspective. In J. A. Katz, & A. C. Corbet (Eds.), *Seminal ideas for the next twenty-five years of advances* (*Advances in entrepreneurship, firm emergence and growth, Vol. 21*) (pp. 141–161). Emerald Publishing Limited.

Mezirow, J. (1991). *Transformative dimensions of adult learning.* Jossey-Bass.

Mezirow, J. (1997). Transformative learning: Theory to practice. *New Directions for Adult and Continuing Education, 74*, 5–12. https://doi.org/10.1002/ace.7401

Ndlela, M. N., Hole, Å. S., Slettli, V. K., Haave, H., Mei, X. Y., Lundesgaard, D., & Namdar, K. (2019). Facilitation of learning in transformative learning circles: Enabling entrepreneurial mindsets through co-creation of knowledge. In D. Higgins, P. Jones, & P. McGowan (Eds.),

Creating entrepreneurial space: Talking through multi-voices, reflections on emerging debates. Emerald Publishing Limited.

Ndofirepi, T. M. (2020). Relationship between entrepreneurship education and entrepreneurial goal intentions: Psychological traits as mediators. *Journal of Innovation and Entrepreneurship, 9*(1), 1–20.

Neergaard, H., Robinson, S., & Jones, S. (2020). Transformative learning in the entrepreneurship education process: The role of pedagogical nudging and reflection. *International Journal of Entrepreneurial Behavior & Research, 27*(1), 251–277.

Nobles, C., & Burrell, D. (2018, March). Using Cybersecurity Communities of Practice (CoP) to support small and medium businesses. In D. A. D. Bedford & Elias G. Carayannis (Eds.), *ICIE 2018 6th International Conference on Innovation and Entrepreneurship: ICIE 2018.* Conference proceedings of Academic Conferences and publishing limited, Washington, DC, USA. https://www.academic-bookshop.com/ourshop/prod_6343438-ICIE-2018-PDF-Proceedings-of-the-6th-International-Conference-on-Innovation-and-Entrepreneurship.html

Nyamunda, J., & Van Der Westhuizen, T. (2020). Developing entrepreneurial self-efficacy: A transformative learning theory approach. *Journal of Contemporary Management, 17*(se1), 44–66.

Polbitsyn, S., Kakouris, A., Kliuev, A., & Bagirova, A. (2020, September). Entrepreneurship education for the digital generation: Invention or transformation? In A. D. Nisco (Ed.), *15th European Conference on Innovation and Entrepreneurship, ECIE 2020.* Conference proceedings of Academic Conferences and Publishing International Limited, Rome, Italy. https://www.academic-bookshop.com/ourshop/prod_7401592-ECIE-2020-PDF-Proceedings-of-the-15th-European-Conference-on-Innovation-and-Entrepreneurship.html

Radipere, S. (2012). South African university entrepreneurship education. *African Journal of Business Management, 6*(44), 11015–11022.

Rasiah, R., Somasundram, S., & Tee, K. P. (2019). Entrepreneurship in education: Innovations in higher education to promote experiential learning and develop future ready entrepreneurial graduates. *Journal of Engineering Science and Technology*, special issue, 99–110.

Rostamian, F. (2019). *Entrepreneurship and new product development syllabus.* https://r.summer.ucla.edu/pdf/2018_TLSI_Syllabus_ProductDevelopment.pdf

Scott, J. M., Pavlovich, K., Thompson, J. L., & Penaluna, A. (2020). Constructive (mis) alignment in team-based experiential entrepreneurship education. *Education + Training, 62*(2), 184–198.

Shane, S. (2010). *Born entrepreneurs, born leaders: How your genes affect your work life.* Oxford University Press.

Stevenson, H. H., & Gumpert, D. E. (1985). The heart of entrepreneurship. *Harvard Business Review, 63*(2), 85–94.

Terzieva, L. (2016). Entrepreneurial learning and communities of practice: The case of the cross-border cultural tourism development Bulgaria-Romania. In J. Ateljević & J. Trivić (Eds.), *Economic development and entrepreneurship in transition economies* (pp. 271–286). Springer.

Timmer, D. (2015). *How to stimulate transformative learning? An explorative study into the triggers of transformative learning in the context of entrepreneurial development* (Bachelor's thesis, University of Twente, Enschede, Netherland). https://essay.utwente.nl/68859/1/BachelorAssignment_DoraTimmer_ResearchReport.pdf

UCLA Extension. (2018). *Entrepreneurship and new venture formation syllabus.* https://uclaextension.campusconcourse.com/view_syllabus?course_id=19822

UCLA: Undergraduate Education Initiatives. (n.d.-a). *Minor requirements: Entrepreneurship.* Retrieved May 20, 2022 from https://www.uei.ucla.edu/academic-programs/entrepreneurship/minor-requirements/

Vettraino, E., & Jones, C. (2022). On the matter of evidencing transformative learning in enterprise/entrepreneurship education. In B. Urzelai & E. Vettraino (Eds.), *Team academy in practice* (pp. 6–20). Routledge.

Wasim, J. (2019). *Closing the gap between university curriculum in entrepreneurship and entrepreneurial learning in networks: An interpretivist constructivist approach* (Doctoral dissertation). University of Portsmouth. https://researchportal.port.ac.uk/en/studentTheses/closing-the-gap-between-university-curriculum-in-entrepreneurship

Weber, R. (2013). *Evaluating entrepreneurship education* (Doctoral dissertation). Universität München. https://doi.org/10.1007/978-3-8349-3654-7

Wei, X., Liu, X., & Sha, J. (2019). How does the entrepreneurship education influence the students' innovation? Testing on the multiple mediation model. *Frontiers in psychology*, *10*, Article 1557.

Weller, A. (2020). Exploring practitioners' meaning of "ethics", "compliance", and "corporate social responsibility" practices: A communities of practice perspective. *Business & Society, 59*(3), 518–544.

Wenger, E. (1998). *Communities of practice: Learning, meaning, and identity*. Cambridge University Press.

Chapter 10
Entrepreneurship Education from a Global Perspective: Successful Experience, Differentiation, and the Way Forward

Xiaozhou Xu and Weihui Mei

The emergence of 4th Industrial Revolution, the global agenda for sustainable development and the increasingly uncertain era have called for the development of innovation and entrepreneurship education. In the past three decades, with the strong promotion of international organizations, governments, enterprises, foundations, and other stakeholders, global entrepreneurship education (EE) in higher education institutions (HEIs) has witnessed tremendous development. Focusing on the practice of EE in eight case countries, this chapter analyzes the successful experience and differentiation of global EE, and highlights its main development trend.

10.1 Successful Experience of EE

It is noted that global entrepreneurship education shares the following successful experience.

10.1.1 From Focused to University-Wide EE

University-wide EE, also known as cross-campus EE, refers to the move of EE programs beyond business schools and its further integration with other academic disciplines (Mei & Symaco, 2022). A university-wide concept abandons the notion that entrepreneurship is just for business students; instead, it applies to administrators, faculty members, and students from any and all parts of the campus (Morris et al., 2013). University-wide EE is seen as a transformational, multidimensional force

X. Xu (✉) · W. Mei
College of Education, Zhejiang University, Hangzhou, China
e-mail: xxiaozhou@163.com

X. Xu, *Comparative Entrepreneurship Education*,
https://doi.org/10.1007/978-981-99-1835-5_10

that enables universities to be more innovative and proactive. Such an approach also further institutionalizes entrepreneurship into HEIs, making it part of its culture and operational model (Morris et al., 2013). Entrepreneurship is also broadly regarded as one of the key competencies for lifelong learning, which further espouses critical thinking, systems thinking, and cultural agility—skills needed to help learners become adaptable in the context of fast technological growth (Aoun, 2017). But beyond simply building businesses, the characteristics of "seeking opportunities, taking risks beyond security, and having the tenacity to push an idea through to reality" marks an entrepreneurial perspective (Kuratko, 2005). Peter Drucker points out: "the entrepreneurial mystique? [...] it is a discipline. And like any discipline, it can be learned" (Kuratko, 2005).

University-wide EE programs have grown over the past decade. For instance, in the US, EE courses are offered to faculties beyond business schools, such as psychology, geography, earth, environmental science, fine and performing arts, and music (Sá & Kretz, 2015). In the UK, the Quality Assurance Agency for Higher Education's (QAA) *Enterprise and Entrepreneurship Education: Guidance for UK Higher Education Providers* (2018) stressed the need for educators to embed entrepreneurship across the curricula. Meanwhile, the *Oslo Agenda for Entrepreneurship Education in Europe* called for better integration across subject areas (European Commission, 2006). Parallel to the development of university-wide EE programs in Western countries, countries in developing regions have also incorporated this program into their HEIs. A common curriculum for entrepreneurship in the Association of Southeast Asian Nations (ASEAN) was proposed to promote better mobility among students, which was seen to complement the development of the region (AsiaSEED, 2012).

However, due the silos in HEIs, limited resources, and the influence of academic evaluation which pays special attention to teaching and research, university-wide EE also faces great challenges (Mei & Symaco, 2022). In order to sustain university-wide EE, Morris et al. (2013) proposed an integrative model to overcome these challenges so as to achieve a truly university-wide approach. The model is built around 12 critical building blocks: academic champion, a common definition, a clear purpose, structure, supporting infrastructure, curricular model, cocurricular programming, resource model, incentives, shared learning, proactive publicity, outcomes and metrics. Though different universities may have different approaches regarding these 12 critical blocks, each university leader should have clear explanations about these 12 elements within the campus so as to effectively promote university-wide EE.

10.1.2 Involvement of Different Stakeholders

Stakeholders are those groups and individuals who can affect or be affected by EE (Freeman, 2010). Bischoff et al. (2018) highlight the importance of stakeholder collaboration within the entrepreneurial ecosystem at HEIs, based on a cross-case analysis of 20 HEIs from 19 European countries. International organizations, national government, enterprises, HEIs, and foundations are the main stakeholders of EE. By implementing different roles and promoting the collaboration of these stakeholders, EE in both developed and developing countries has grown rapidly.

International organizations play an important role in disseminating EE ideas, organizing academic conferences, and formulating EE standards, etc. For example, since 2011, UNESCO Bangkok has been involved in organizing the series of UNESCO-APEID (Asia and the Pacific Programme of Educational Innovation for Development) meetings in EE with various countries. This highlights the growing role of EE in countries beyond the West and the greater involvement of international agencies in the process (UNESCO, 2017).

Governments play a crucial role in improving the institutional environment for the development of EE by formulating specific EE policies, or integrating related policies into more comprehensive innovation policies or higher education policies.

HEIs are the main force of EE reform. On the one hand, HEIs deepen the "student-centered" teaching idea and strengthen the knowledge and skills to meet the needs of the changing marketplace; on the other hand, HEIs actively promote the integration of EE into different disciplines so as to cultivate professional talents with entrepreneurial mindsets and abilities.

Enterprises are also active supporters of EE. They not only provide practice platforms, financial support, personnel resources for HEIs, but are also directly involved in curricula, co-curricula and other entrepreneurial practices to introduce the real world into campuses.

Additionally, philanthropic foundations are important stakeholders for EE. For example, the Kauffman Foundation in the US has implemented two rounds of Kauffman Campus Initiative (KCI), cumulatively providing USD45 million, plus another USD148 million of matching funds, to promote cross-campus EE (Morris et al., 2013; Schneider, 2015, p. 15). The selection criteria for the KCI funds are as follows: (1) whether a school could make entrepreneurship a common and accessible activity for all students; (2) the level of involvement of the president or chancellor; (3) the ability to generate matching funds; (4) whether the school could serve as a model for other colleges and universities; (5) the relative strength of the innovative approaches; and (6) the likelihood that the initiative would change campus culture and produce a sustainable entrepreneurial spirit on campus (Morris et al., 2013, p. 246).

10.1.3 Attaching Importance to Curricular and Extracurricular Development

The starting point of EE in some countries was a single for-credit course (e.g., the US), while in other countries it was introduced through extracurricular activities like business plan competitions (e.g., China). Subsequently, both curricular and extracurricular EE offerings developed rapidly such that both serve diverse students from different academic backgrounds. The case countries have gradually implemented an increasing number of entrepreneurship education programs.

Traditional entrepreneurship courses were mostly provided by schools of business, which are designed based on three main approaches: core elements constituting a business plan, steps in the entrepreneurial process, and the life cycle of a venture (Morris et al., 2013, pp. 59–60).In the US, Introduction to Entrepreneurship, Business Planning, and Entrepreneurial Financing are the most popular courses (Center for Entrepreneurial Excellence, 2014). In Finland, EE is widely offered in Finnish universities at least in some form of individual entrepreneurship courses, which implies that tremendous development took place from 2005 to 2016.

With the rising importance of innovation and entrepreneurship, along with the rise of university-wide EE, the curriculum moved beyond business schools. Interdisciplinary courses were developed and open to students from across campus interested in entrepreneurship (Morris et al., 2013, pp. 73–74). Under such circumstance, there was a boom in entrepreneurship curriculum. With the introduction of courses like Green Entrepreneurship, Social Entrepreneurship, Digital Entrepreneurship, Technology Entrepreneurship, Entrepreneurship and Education, Entrepreneurship and Psychology, etc., the EE curriculum has become increasingly diversified.

Besides, due the practicality of EE, HEIs have also attached great importance to providing different types of project-based extracurricular activities such as simulations, elevator pitches, business plan competitions, entrepreneurial internships, student venture incubators, etc. At the same time, it is also pertinent that HEIs establish a network of entrepreneurs, angel investors, venture capitalists, lawyers, which can provide information, resources, guidance, and financial support for students and faculty interested in entrepreneurship.

10.1.4 Establishment of the EE Ecosystem

The term *ecosystem* has been used to explain the succession, competition, and metabolism not only in the natural system, but also that in social and human systems (Brush, 2014; Christopherson, 1997). In the last decade, with the rise of the entrepreneurial economy and the urgent demand for entrepreneurial talents, the concept of establishing entrepreneurship ecosystems has become a popular topic around the world (Brush, 2014). An ecosystem is the interaction of people, roles, infrastructure, organizations, and events which creates an environment for heightened levels of entrepreneurial activity (Neck et al., 2004; Regele & Neck, 2012). A McKinsey report (Mckinsey & Company, 2011) emphasized that shaping fertile ecosystems, financing new ventures from inception to critical size, and infusing the population with an entrepreneurial culture are the three pillars for boosting the entrepreneurial engine. The efforts to create an entrepreneurial ecosystem show a nation's determination in the long-term development of entrepreneurship activities, not the short-term solutions.

Given the importance of the entrepreneurship ecosystem, many international organizations and researchers have conducted deep analyses of the domains of an entrepreneurship ecosystem (see Table 10.1). Isenberg (2010) presents six domains of an entrepreneurship ecosystem: conducive culture, enabling policies and leadership, availability of finance, quality human capital, venture-friendly markets for products, and a range of institutional and infrastructural support. Isenberg also points out that the individual elements combine in complex ways, and in isolation; each is conducive to entrepreneurship, but insufficient to sustain it. This means that an entrepreneurship ecosystem requires each element to coordinate with each other and to maintain a dynamic balance. Another important point raised by Isenberg (2010) is that many governments take a misguided approach to building entrepreneurship ecosystems by pursing some unattainable ideal of an ecosystem and look to economics that are completely unlike theirs for best practices. To establish an entrepreneurship ecosystem, we should seriously consider the local conditions.

The EE ecosystem is a sub-ecosystem within the broader entrepreneurship ecosystem (Neck & Liu, 2021; Regele & Neck, 2012). Different stakeholders play different roles in the ecosystem (WEF, 2009; Xu, 2019). HEIs play a critical role as intellectual hubs in the entrepreneurial ecosystem by serving as an incubator for innovation and research as well as a focal network for collaboration among researchers, students, professors, companies, venture capital firms, angel investors, and entrepreneurs (WEF, 2009). The main activities of an internal EE ecosystem include curriculum, cocurricular activities, and research, while the necessary support system comprises stakeholders, resources, infrastructure, and culture (Fetters et al., 2010; Neck & Liu, 2021).

Table 10.1 Key pillars of entrepreneurial ecosystems

Isenberg (2010)	GEM (2022)	WEF (2014)
Conducive culture Enabling policies and leadership Availability of finance Quality human capital Venture-friendly markets for products Institutional and infrastructural supports	Entrepreneurial Finance Government Policies Support and Relevance Government Policies: Taxes and Bureaucracy Government Entrepreneurship Programs Entrepreneurial Education at School Stage Entrepreneurial Education at Post-school Stage R&D Transfer Commercial and Legal Infrastructure Internal Market: Dynamics Internal Market: Burdens or Entry Regulation Physical Infrastructure Cultural and Social Norms	Accessible markets Funding & finance Government & regulatory framework Major universities as catalysts Human Capital/workforce Support systems/mentors Education & training Cultural support

10.2 Differentiation of EE

As discussed above, the development of EE has shown an obvious tendency of convergence. However, affected by the political system, economic development stage, cultural background, and other factors, there are also significant differences in the development of EE among countries, and even among regions of the same country. It should be noted that EE is closely related to the local support conditions, talent training needs, industrial structures, etc. We should also pay attention to the localization of EE. The following section will analyze the differentiation of EE from the four dimensions of objectives and content, developmental model, institutional environment, and ways of evaluation.

10.2.1 Differentiation of Objectives and Content

Different understandings to EE may lead to different objectives and contexts. Some studies argued that there are three levels of EE objectives: learning to understand entrepreneurship, learning to become entrepreneurial, and learning to become an entrepreneur (Hytti & O'Groman, 2004). With the fast development of university-wide EE (Mei & Symaco, 2022; Morris et al., 2013), how to make the content and the delivery of EE suitable for students with different needs is an important question. The UK's QAA (2018) guidelines identify and distinguish three ways in which enterprise and EE can be delivered: learning "about" entrepreneurship, learning "for" entrepreneurship, and learning "through" entrepreneurship. These differentiations call for open discussions about the following questions (Kuratko & Hoskinson, 2017). When delivering EE, should we adopt the lecture format or experiential teaching methods? Which is more effective—the business plan or the lean startup approach? Should we develop entrepreneurship majors or integrate entrepreneurship into different disciplines? How can we promote HEI–enterprise collaboration? These discussions are important for the further development of EE.

10.2.2 Differentiation of Developmental Models

The differentiation of developmental models can be observed through both external and internal perspectives. Externally, there are two main types of EE developmental models: the centralized and decentralized. The centralized developmental model is a top-down approach guided by the central government's policies and other incentives. For example, China issued *Implementation Opinions on Deepening the Reform of Innovation and Entrepreneurship Education in Higher Education Institution* (2015) and *Guidance on Further Supporting College Students' Innovation and Entrepreneurship* (2021) to promote EE. China's Ministry of Education has also

Table 10.2 Comparison of magnet and radiant models of university-wide EE

Magnet model benefits	Radiant model benefits
• An easier model to maintain, given the one-stop approach to EE • Greater research benefits to the business school when a critical mass of entrepreneurship faculty is developed • An identifiable program on campus for alumni to give to • An academically diverse group of students in entrepreneurship classes • Loyalty to the school of business from students across campus	• A broader reach to students who can more easily find a course with entrepreneurship in their major • Context-specific training in law, art, engineering, history, or other disciplines for non-business students • Collaboration among faculty and students across academic units • A broader group of alumni entrepreneurs that did not graduate from the school of business Loyalty to the university across disciplines

Source Antal et al. (2014, p. 230)

recognized 200 HEIs as having "Typical Experience of Entrepreneurship Education" since 2016, and these set examples for other HEIs (MOE, 2019). The UK, South Korea, and Finland follow the same developmental models. The decentralized developmental model is a bottom-up approach that emphasizes the importance of local government or the initiatives of each HEI. Canada, the US, and Germany follow the decentralized developmental model of EE.

Internally, although university-wide EE has been widely accepted globally, there are still differentiated means of implementation. The international experience shows that there are two main structural models of university-wide EE: magnet and radiant (Streeter et al., 2011). In the magnet model, one college is responsible for the EE in HEIs, while the radiant model encourages different colleges to be involved in EE. A comparison of the two models is illustrated in Table 10.2 (Antal et al., 2014, p. 230). HEIs can implement the appropriate model based on their historical background, resources, and opinions of faculty and students, so as to deliver high-quality EE.

10.2.3 Differentiation of Institutional Environment

Based on the data of Global Entrepreneurship Monitor (2021/2022), this study makes a comparative analysis of the institutional environment of the eight case countries (Table 10.3 and Fig. 10.1). Of the eight countries, Finland scores highest in many indicators, which shows that Finland has established a good entrepreneurship ecosystem in terms of a robust entrepreneurial finance sector, R&D transfer system, commercial and legal infrastructure, as well as physical infrastructure, etc. However, GEM's 2021 Entrepreneurial Activity Review found that despite the high quality of institutional environment for entrepreneurship, the Finnish population seemed discouraged when considering starting a business (GEM, 2022, p. 111). This is evident from the low rate of Finnish population who say they have the knowledge, skills, and experience required to start a business (42.8%), and for those who expect to start a business in the next three years (9.8%) (GEM, 2022, p. 111). This phenomenon needs further study. Regarding other case countries, South Korea scores highest in indicators of government policies (support and relevance) and internal market (dynamics); Germany scores highest in indicator of government entrepreneurship programs; and the US scores highest in cultural and social norms (GEM, 2022). Countries should understand their advantages and disadvantages of their institutional environment, so that further reform can be relevant.

In recent years, GEM also introduced the National Entrepreneurship Context Index (NECI) to compare different economy's institutional environment for entrepreneurship and EE. GEM found that those countries who saw consistent and substantial improvements in their NECI scores in recent years were those who mitigated the impact of the pandemic on new startups, or who supported women entrepreneurs (GEM, 2022, p. 91). As the pandemic becomes the new normal, policymakers and HEIs should fully consider the impact of the epidemic on college students' entrepreneurship, or see the Covid-19 pandemic as a potential trigger for entrepreneurship (GUESSS, 2021, p. 24). Besides, special attention should be paid to the gender gap in entrepreneurship and systematically promote female entrepreneurs (GUESSS, 2021, p. 24).

Table 10.3 Comparison of GEM indicators of eight countries

	1	2	3	4	5	6	7	8	9	10	11	12
Canada	5.1	4.6	5.9	4.7	4.0	4.7	4.2	6.0	4.6	5.1	6.6	5.7
China*	5.8	5.9	6.2	5.5	4.1	5.7	5.6	5.4	6.9	5.2	7.7	6.8
Croatia	4.3	2.7	3.4	4.1	2.7	3.4	3.3	4.8	5.8	3.5	6.0	3.0
Finland	7.1	5.5	6.3	5.6	6.1	6.0	6.0	6.9	4.2	6.1	8.6	5.4
Germany	5.3	4.4	4.7	6.4	2.8	5.2	4.9	6.3	5.3	5.0	6.1	4.6
South Korea	5.6	6.4	5.9	5.9	4.3	4.8	4.5	5.0	7.8	4.8	7.7	5.7
UK	5.2	4.2	5.6	4.3	3.2	5.0	4.2	5.8	4.9	5.5	6.5	5.3
USA	6.2	4.2	5.3	4.2	3.2	5.0	4.7	6.4	5.6	4.7	7.5	7.0

Source GEM (2022)
Note 1 Entrepreneurial Finance; 2 Government Policies: Support and Relevance; 3 Government Policies: Taxes and Bureaucracy; 4 Government Entrepreneurship Programs; 5 EE at School Stage; 6 EE at Post-school Stage; 7 R&D Transfer; 8 Commercial and Legal Infrastructure; 9 Internal Market: Dynamics; 10 Internal Market: Burdens or Entry Regulation; 11 Physical Infrastructure; 12 Cultural and Social Norms
*China uses data from 2019–2020

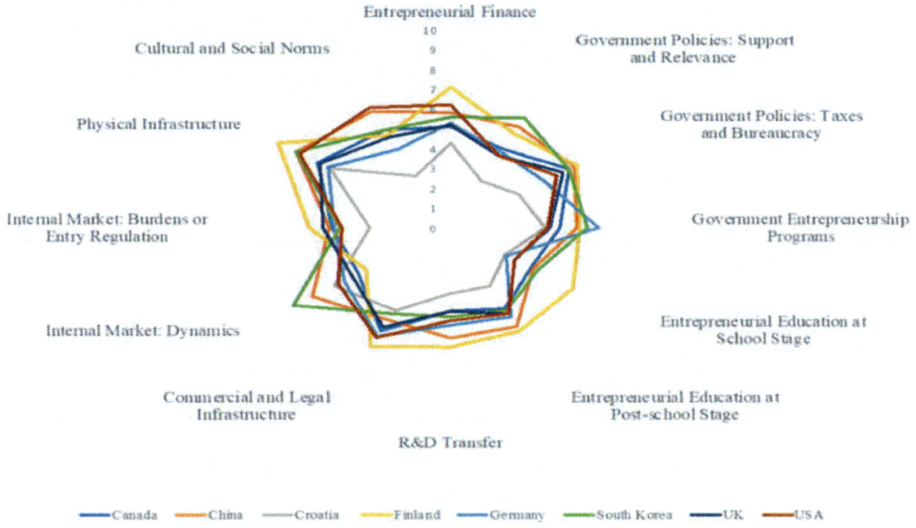

Fig. 10.1 Comparison of GEM indicators of eight countries. *Source* GEM (2022) (China uses data from 2019–2020)

10.2.4 Differentiation of Evaluation

With the boom of EE programs, researchers have identified the importance and necessity of evaluating the effectiveness of EE (Yu & Mei, 2021). Further empirical research studies have proved the active impact of EE on cultivating students' entrepreneurial spirit and enhancing the rate of entrepreneurship (Lee & Eesley, 2018). For example, Lee and Eesley (2018) examined the impact of two entrepreneurship programs at Stanford University, indicating that participation in the Center for Entrepreneurial Studies at the Business School is associated with a 17% higher probability of becoming an entrepreneur, and participation in Stanford Technology Ventures Program at the Engineering School is associated with a 6.1% higher probability (Lee & Eesley, 2018).

However, the evaluation of EE is much more complicated than just exploring the impact of EE on students' probability of becoming an entrepreneur. There are two basic means of evaluation of EE programs: process evaluation and impact evaluation. A process evaluation aims to explain how an EE program is organized and implemented, while an impact evaluation focuses on the outcomes or effectiveness of the EE programs (Yu & Mei, 2021). The evaluation objects include both individual and the organization. This study tries to propose a four-quadrant model for EE evaluation. It is built on two axes (Fig. 10.2). The horizontal axis is the content dimension of evaluation. At one end is Process Evaluation and at the other end is Outcome Evaluation. The vertical axis is the object dimension of evaluation. At the ends are individual evaluation and organizational evaluation. The evaluation indicators of each quadrant are also listed. As illustrated in Fig. 10.2, there are different types of EE evaluation,

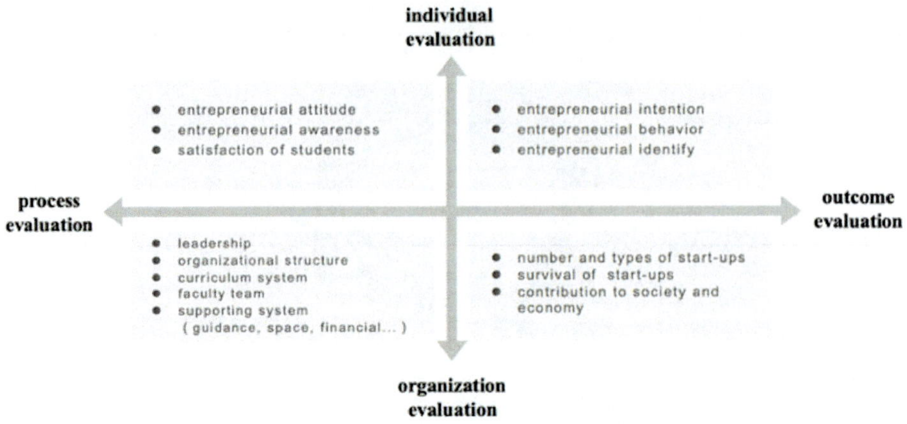

Fig. 10.2 A four-quadrant model for EE evaluation. *Source* made by authors

based on different evaluation purposes. Whether the evaluation is for strategic planning, monitoring the program, or impact assessment, different evaluation approaches may be needed.

Besides the individual and institutional levels of evaluation, some international organizations have also tried promoting the national efforts of supporting innovation and entrepreneurship. For example, the OECD and the European Commission have cooperated to establish the HEInnovate online platform—a free self-assessment tool for all types of HEIs to evaluate their status of innovation and entrepreneurship. Based on HEInnovate, the OECD and the European Commission have published country reports. The first country report, *Supporting Entrepreneurship and Innovation in Higher Education in Ireland*, was published in 2017, followed by *Poland* (2017), *Hungary* (2017), *The Netherlands* (2018), *Austria* (2019), and *Italy* (2019). These reports provide systematic evidence on the efforts of these countries to support innovation and entrepreneurship as well as indicates the characteristics and challenges of different countries.

10.3 The Future of EE

It has been more than 70 years since the first entrepreneurship course was taught in Harvard University. Since then, both the external and internal environment of EE in HEIs have undergone tremendous changes. In 18–20 May 2022, the third World Higher Education Conference (UNESCO, 2022) was held in Barcelona. It identified six principles for the future of higher education: (1) Inclusion, equity, and pluralism; (2) Academic freedom and participation of all stakeholders; (3) Inquiry, critical thinking, and creativity; (4) Integrity and ethics; (5) Commitment to sustainability and social responsibility; (6) Excellence through cooperation rather than competition (UNESCO, 2022). Though these are general principles for the development of higher education as a whole, they are also significant for the future of EE. For example, the future of EE calls for strategic development so that it can provide equitable access for all students. At the same time, it can also stimulate HEIs to make a campus-wide commitment to sustainability and social responsibility by encouraging students to find solutions for global challenges. In the following section, we identify four trends of the future development of EE.

10.3.1 Strategic Development: Systematically Support the Development of EE

In the context of globalization and technology development, innovation and entrepreneurship have become the main proponents for international organizations and countries to promote and sustain development (Mei & Symaco, 2022; UNESCO, 2015). In order to provide students from different academic background with equal opportunity to gain access to EE, it is necessary to make EE an independent strategy or a part of an integrated strategy to cultivate innovative and entrepreneurial talents.

Some international organizations and countries have attached great importance to EE strategy since the early twenty-first century. The European Commission of the EU, for example, regards "enhancing creativity and innovation, including entrepreneurship" as one of the four objectives of the strategic framework for European cooperation in education and training (European Commission, 2009). EU-wide policies have been issued to enhance the culture of entrepreneurship. These include *Entrepreneurship in European Green Paper* (2003), *The Oslo Agenda for Entrepreneurship Education in Europe* (2006), *Towards Greater Coherence in Entrepreneurship Education* (2010), *Entrepreneurship 2020 Action Plan* (2012), *Entrepreneurship Education: A Road to Success* (2015). Affected by these EU policies, European countries are also very active in implementing EE strategies to systematically promote it. Figure 10.3 illustrates the introduction and duration of specific EE strategies implemented in Europe during 2000–2015 (European Commission, 2016, p. 39).

In order to support developing countries' policymakers in the design of their national entrepreneurship strategy, the United Nations Conference on Trade and Development (UNCTAD) issued the Entrepreneurship Policy Framework and Implimentation Guidance. Part of this framework is enhancing EE and skills development, proposing clear policy objectives in five dimensions: embedding entrepreneurship in formal and informal education; developing effective entrepreneurship curricula; training teachers; strengthening the insitutional framework; and partnering with the private sector (UNCTAD, 2022). According to this framework, developing countries can formulate national innovation and entrepreneurship strategies based on their actual situation.

The development of social EE has also received increasing attention internationally. For example, the *Strategy for Development of Social Entrepreneurship in the Republic of Croatia 2015–2020* highlights the importance of social EE and puts forth

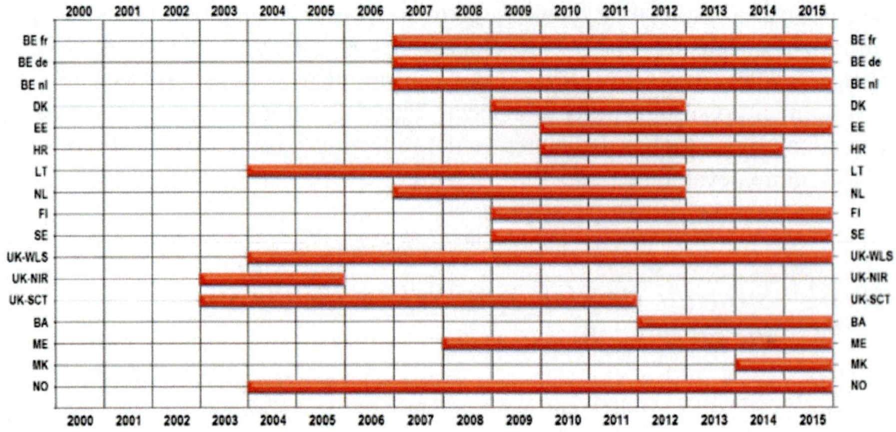

Fig. 10.3 Implementation of specific central-level EE strategies, 2000–2015

the following objectives: (1) establishment and improvement of the legislative and institutional framework for the development of social entrepreneurship; (2) establishment of a financial framework for the effective performance of social entrepreneurs; (3) promoting the importance and role of social entrepreneurship through all forms of education; and (4) ensuring the visibility of the role and possibilities of social entrepreneurship in the Republic of Croatia and informing the general public about themes related to social entrepreneurship issues (European Commission, 2015).

At the institutional level, many universities have also adopted a strategic approach to promote the development of EE. Some universities have a comprehensive university-wide strategy dedicated to EE. For example, University College London (UCL) published *UCL Innovation and Enterprise Strategy 2016–2021: Transforming Knowledge and Ideas into Action*, which served as "an approach and framework to create a real spirit of enterprise at UCL." The University of Queensland's (UQ) *Entrepreneurship Strategy 2018–2022* proposes that "every student will have the opportunity to experience entrepreneurial learning at UQ." Other universities incorporate promoting EE into the university strategy. For instance, the University of Waterloo included the goal, "Propel Waterloo's global leadership in *innovation, entrepreneurship* and social impact," in its university strategic plan 2020–2025. One of UC Berkeley's strategies is "Berkeley embraces the California spirit: diverse, inclusive, *entrepreneurial*."

10.3.2 Institutionalization: Reforming Organizational Structures and Supporting Institutions

As the historian Cobban (1975) said, "The absence of regular organization may initially provide a fillip for free-ranging inquiry, but perpetuation and controlled development can only be gained through an institutional framework." Similarly, the sustainable development of EE also to a great extent depends on the level of its institutionalization.

With more and more students interested in entrepreneurship, universities tend to adopt a cross-campus and interdisciplinary approach to extend EE to all students regardless of their academic background, which calls for the reform of organizational structures and supporting institutions. In the US, entrepreneurship centers, departments of management and entrepreneurship, departments of entrepreneurship, and schools of entrepreneurship are the most common structural forms to deliver entrepreneurship education (Morris et al., 2013). In China, colleges of entrepreneurship are the dominant structure for cross-campus EE in HEIs (Mei & Symaco, 2022).

Besides academic structures, startup space infrastructures, such as incubators and accelerators, have also been established to support faculty or student entrepreneurship. For example, the University of Toronto entrepreneurship community consists of more than 10 accelerators across three campuses, and it is now the top-ranking university for research-based startups in Canada. Meanwhile, Finland's Aalto University aims to provide every student and staff member with experience of entrepreneurial thinking and action (Reichert, 2019). The Aalto University Startup Center provides both a pre-incubator program and an accelerator program. The former focuses on projects that are at the idea phase and in technological research, and the latter focuses on sustainable research-based and innovative deep-tech startups.

Universities need to become more entrepreneurial to better conduct their strategy and to make EE more institutionalized. The concept of "entrepreneurial university" (Clark, 1998; Etzkowitz, 2003) has therefore attracted great attention from the government and universities. In the UK, the National Centre for Entrepreneurship in Education (NCEE) supports HEIs to develop their entrepreneurial capacity. Since 2008, the NCEE has sponsored the Outstanding Entrepreneurial University Award to encourage an entrepreneurial culture throughout the UK. The past five years' winners are Sheffield Hallam University (2021), Aston University (2020), Loughborough University (2019), King's College London (2018), and Liverpool John Moores University (2017).

10.3.3 Professionalization: Making EE an Academic Area of Study

EE comprises practical and theoretical aspects. Besides the vigorous and popular entrepreneurial activities, it is also of great importance to make EE an academic area of study and to conduct high-quality entrepreneurship education research. The latter includes exploring the rules of college students' entrepreneurship, providing empirical evidence for the impact of EE, studying the effectiveness of national youth entrepreneurship policies, etc. International experience shows that only by consistently producing high-quality research output that is of comparable quality to that of other academic fields can it gain legitimacy in the university setting.

The field of EE has grown exponentially in the past decades. On the one hand, this field has become one of the most fast-developing fields, with the prolific publication of monographs, academic journals. Tiberius and Weyland (2022) made bibliometric analyses of 680 articles with the term "entrep* education*" on the Web of Science (WoS), and found that *Education + Training* (56), *Industry of Higher Education* (26), *Annals in Entrepreneurship Education* (22), *Journal of Small Business Management* (21), *Frontiers in Psychology* (19) are the top five journals that publish EE-related articles. *Studies in Higher Education* (11), *Sustainability* (11), *Journal of Business Venturing* (8), and *Journal of Technology Transfer* (8) also feature EE-related articles. It should be noted that both the entrepreneurship journals and higher education journals are increasingly interested in EE articles.

On the other hand, there are more discussions around curriculum, teaching methods, entrepreneurship practice, and so on. Due to the importance of entrepreneurship to both individual and social development, the debate on "whether entrepreneurship can be taught" has gradually turned to "what to teach" and "how to teach." There are five levels of learning in EE (Johannisson, 1991; Souitaris et al., 2007): know-why (values and motivation of entrepreneurs), know-what (knowledge about what needs to be done), know-how (practical abilities and skills), know-who (awareness of social networks and the ability to use them), and know-when (experience and intuition about when to take action). Regarding how to teach, Nabi et al. (2017) systematically reviewed 159 published articles and found that there were four teaching models of entrepreneurship. These are: the supply model focusing on reproduction methods such as lectures, reading, etc.; the demand model focusing on personalized/participative methods such as simulations; the competence model focusing on communication, discussion, and production methods; and the hybrid model. Different teaching models may have different impact on students' entrepreneurial intention and behaviors.

Another trend is that the Covid-19 pandemic and the advancement of digital technology have accelerated the integration of technology and EE globally (Hyams-Ssekasi & Yasin, 2022). On the one hand, the use of Massive Open Online Courses (MOOCs) has grown in popularity. Platforms like Coursera, edX, Khan Academy, Udacity, FutureLearn, OpenupEd, and Iersity, provide free online entrepreneurship courses for learners from all over the world. In China, more than 460 online entrepreneurship courses can be found on the icourse platform.[1] On the other hand, the innovation of learning technologies, such as augmented reality, virtual reality, and artificial intelligence, are starting to have an influence on the reform of EE through blended, hybrid, and rotational models of learning design (Hyams-Ssekasi & Yasin, 2022).

---

[1] https://www.icourse163.org.

10.3.4 Internationalization: Strengthening Global EE Exchange and Cooperation

In a global connected world, cultivating innovative and entrepreneurial talents with a global vision who are good at identifying global entrepreneurial opportunities and responding to global challenges is one of the most important objectives of EE in colleges and universities. Students with global competence demonstrate four competencies: (1) investigate the world beyond their immediate environment, framing significant problems and conducting well-crafted and age-appropriate research; (2) recognize perspectives, others' and their own, articulating and explaining such perspectives thoughtfully and respectfully; (3) communicate ideas effectively with diverse audiences, bridging geographic, linguistic, ideological, and cultural barriers; (4) take action to improve conditions, viewing themselves as players in the world and participating reflectively (Mansilla & Jackson, 2011). Since 2018, the OECD has been conducting the global competence assessment, which uses the following definition, "Global competence is the capacity to examine local, global and intercultural issues, to understand and appreciate the perspectives and world views of others, to engage in open, appropriate and effective interactions with people from different cultures, and to act for collective well-being and sustainable development" (OECD, 2019).

International organizations play a significant role in promoting global exchange and cooperation of EE. For example, by working with ministries of education, HEIs, international organizations, and private sector entities, UNESCO Bangkok established the regional Entrepreneurship Education Network (EE-Net) to conduct comparative research on EE and to create a database of successful entrepreneurs in Asia and the Pacific to serve as mentors for future entrepreneurs (UNESCO Bangkok Office, n.d.).

HEIs have also been actively promoting the cooperation and communication of EE. For example, the Global Consortium of Entrepreneurship Centers (formerly known as the National Consortium of Entrepreneurship Centers in the US) was established in 1997 and now includes 250 university-based entrepreneurship programs throughout the world. In Finland, the college student startup event Slush has grown from a single gathering in Helsinki to a series of events organized all around the world, which creates and supports the next generation of groundbreaking entrepreneurs (Slush, 2022).

In the post-pandemic era, facilitated by advanced digital technologies, an abundance of global entrepreneurship opportunities will emerge. Therefore, various international exchange and cooperation of EE should be enhanced in the future.

References

Aalto University Startup Center. (2022). *Services-accelerator program*. Retrieved from https://startupcenter.aalto.fi/services/accelerator-program/

Antal, N., Kingma, B., Moore, D., & Streeter, D. (2014). University-wide entrepreneurship education. In S. Hoskinson & D. F. Kuratko (Eds.), *Innovative pathways for university entrepreneurship in the 21st century* (pp. 227–254). Emerald Group Publishing Limited.

Aoun, J. E. (2017). *Robot-Proof: Higher education in the age of artificial intelligence*. The MIT Press.

AsiaSEED. (2012). *Common curriculum for entrepreneurship in ASEAN*. https://asean.org/wp-content/uploads/2012/05/8.-Common-Curriculum-for-Entrepreneurship-in-ASEAN.pdf.

Bischoff, K., Volkmann, C. K., & Audretsch, D. B. (2018). Stakeholder collaboration in entrepreneurship education: An analysis of the entrepreneurial ecosystems of European higher education institutions. *Journal of Technology Transfer, 43*, 20–46.

Brush, C. G. (2014). Exploring the concept of an entrepreneurship education ecosystem. In S. Hoskinson & D. F. Kuratko (Eds.), *Innovative pathways for university entrepreneurship in the 21st century* (pp. 25–39). Emerald Group Publishing Limited.

Center for Entrepreneurial Excellence. (2014). *The national survey of entrepreneurship education: An overview of 2012–2014 survey data*. Retrieved from http://www.nationalsurvey.org/files/2014KauffmanReport_Clean.pdf

Christopherson, R. W. (1997). *Geosystems: An introduction to physical geography*. Prentice Hall.

Clark, B. (1998). *Creating entrepreneurial universities: Organizational pathways of transformation*. Pergamon Press.

Cobban, A. B. (1975). *The medieval universities: Their development and organization*. Methuen & Co Ltd.

Etzkowitz, H. (2003). Research Group as "quasi-firms": The invention of the entrepreneurial university. *Research Policy, 32*(1), 109–121.

European Commission. (2006). *The Oslo agenda for entrepreneurship education in Europe*. Retrieved from https://ec.europa.eu/docsroom/documents/8968/attachments/1/translations/en/renditions/pdf

European Commission. (2015). *The strategy for development of social entrepreneurship in the Republic of Croatia 2015–2020*. https://betterentrepreneurship.eu/en/node/33.

European Commission. (2016). *Entrepreneurship education at school in Europe* (Eurydice Report). Publications Office of the European Union.

European Commission. (2009). *A strategic framework for European cooperation in education and training ('ET 2020')*. Retrieved from https://eur-lex.europa.eu/LexUriServ/LexUriServ.do?uri=OJ:C:2009:119:0002:0010:EN:PDF.

Fetters, M., Greene, P., Rice, M., & Bulter, J. S. (2010). *The development of university-based entrepreneurship ecosystems: Global perspectives*. Edward Elgar Publishing.

Freeman, R. E. (2010). *Strategic management: A stakeholder approach*. Cambridge University Press.

Global Consortium of Entrepreneurship Centers. (n.d.). *Mission*. Retrieved from http://www.globalentrepreneurshipconsortium.org/mission/

Global Entrepreneurship Monitor. (2022). *Global entrepreneurship monitor 2021/2022 global report: Opportunity amid disruption*. Retrieved from https://gemconsortium.org/report/gem-20212022-global-report-opportunity-amid-disruption

GUESSS (Global University Entrepreneurial Spirit Students' Survey). (2021). *Global students entrepreneurship 2021: Insights from 58 countries: 2021 GUESSS global report*. Retrieved from https://www.guesssurvey.org/resources/PDF_InterReports/GUESSS_2021_Global_Report.pdf

Hyams-Ssekasi, D., & Yasin, N. (2022). *Technology and entrepreneurship education: Adopting creative digital approaches to learning and teaching*. Palgrave Macmillan.

Hytti, U., & O'Gorman, C. (2004). What is "enterprise educations"? An analysis of the objectives and methods of enterprise education programs in four European countries. *Education + Training, 46*(1), 11–23.

Isenberg, D. J. (2010, June). How to start an entrepreneurial revolution. *Harvard Business Review*, 2–11.

Johannisson, B. (1991). University training for entrepreneurship: Swedish approaches. *Entrepreneurship & Regional Development, 3*(1), 67–82.

Kuratko, D. F. (2005). The emergence of entrepreneurship education: Development, trends, and challenges. *Entrepreneurship Theory and Practice, 29*(5), 577–598.

Kuratko, D. F., & Hoskinson, S. (2017). *The great debates in entrepreneurship.* Emerald Publishing.

Lee, Y. S., & Eesley, C. (2018). The persistence of entrepreneurship and innovative immigrants. *Research Policy, 47*(6), 1032–1044.

Mansilla, V. B., & Jackson, A. (2011). *Educating for global competence: Preparing our youth to engage the world.* Asia Society & Council of Chief State School Officers. Retrieved from https://asiasociety.org/files/book-globalcompetence.pdf

McKinsey & Company. (2011). *The power of many: Realizing the socioeconomic potential of entrepreneurs in the 21st century economy.* Retrieved from https://www.citizen-entrepreneurs.com/wp-content/uploads/2018/04/McKinseyCompany-ThePowerofMany-ACalltoAction-June-7th-2012.pdf

Mei, W. H., & Symaco, L. (2022). University-wide entrepreneurship education in China's higher education institutions: Issues and challenges. *Studies in Higher Education, 47*(1), 177–193.

Ministry of Education of the People's Republic of China. (2019). *Notice on publishing the list of HEIs with typical experience of entrepreneurship education in 2019.* Retrieved from http://www.moe.gov.cn/srcsite/A15/s7063/201908/t20190822_395517.html [In Chinese]

Morris, N. M., Kuratko, D. F., & Pryor, C. G. (2013). Building blocks for the development of university-wide entrepreneurship. *Entrepreneurship Research Journal, 4*(1), 45–68.

Nabi, G., et al. (2017). The impact of entrepreneurship education in higher education: A systematic review and research agenda. *Academy of Management Learning & Education, 16*(2), 277–299.

Neck, H. M., & Liu, Y. P. (2021). *Innovation in global entrepreneurship education: Teaching entrepreneurship in practice.* Edward Elgar Publishing.

Neck, H. M., Meyer, G. D., Cohen, B., & Corbett, A. C. (2004). An entrepreneurial system view of new venture creation. *Journal of Small Business Management, 42*(2), 190–208.

Organization for Economic Cooperation and Development. (2019). PISA 2018 global competence framework. In *PISA 2018 Assessment and Analytical Framework* (pp. 165–216). OECD Publishing. Retrieved from https://doi.org/10.1787/043fc3b0-en

Quality Assurance Agency for Higher Education. (2018, January). *Enterprise and entrepreneurship education: Guidance for UK higher education providers.* Retrieved from https://www.qaa.ac.uk/docs/qaas/enhancement-and-development/enterprise-and-entrpreneurship-education-2018.pdf?sfvrsn=15f1f981_8

Regele, M. D., & Neck, H. M. (2012). The entrepreneurship education sub-ecosystem in the United States: Opportunities to increase entrepreneurial activity. *Journal of Business & Entrepreneurship, 23*(2), 25–47.

Reichert, S. (2019). *The role of universities in regional innovation ecosystem.* European University Association.

Sá, C. M., & Kretz, A. J. (2015). *The entrepreneurship movement and the university.* Palgrave MacMillan.

Schneider, M. (2015). *Kauffman campus initiative: A study that explores the phenomenon of cross-campus entrepreneurship.* (Doctoral dissertation, University of Pennsylvania, Philadelphia, Pennsylvania, the United States). Retrieved from https://www.proquest.com/dissertations-theses/kauffman-campuses-initiative-study-that-explores/docview/1727125381/se-2.

Slush. (2022). Retrieved from https://www.slush.org/about/

Souitaris, V., Zerbinati, S., & Al-Laham, A. (2007). Do entrepreneurship programmes raise entrepreneurial intention of science and engineering students? The effect of learning, inspiration and resources. *Journal of Business Venturing, 22*(4), 566–591.

Streeter, D. H., Kher, R., & Jaquette, Jr. J. P. (2011). University-wide trends in entrepreneurship education and the rankings: A dilemma. *Journal of Entrepreneurship Education, 14*, 75–92.

Tiberius, V., & Weyland, M. (2022). Entrepreneurship *education* or *entrepreneurship* education? A bibliometric analysis. *Journal of Further and Higher Education*. Retrieved from https://www.tandfonline.com/doi/full/10.1080/0309877X.2022.2100692

UNESCO. (2015). *2030 Sustainable development goals*. Retrieved from https://en.unesco.org/sustainabledevelopmentgoals

UNESCO. (2017). *Entrepreneurship education*. Retrieved from https://bangkok.unesco.org/content/entrepreneurship-education

UNESCO. (2022, May 18–20). *Beyond limits: New ways to reinvent higher education*. Working document proposed for the 3rd World Higher Education Conference, Barcelona, Spain. Retrieved from https://cdn.eventscase.com/www.whec2022.org/uploads/users/699058/uploads/69c2df623079c3845e236c56ba2d7a8aa21b3d75489e28c7910226f24f7989aec7aae05a23f31fae4587aeb4be088f9

UNESCO Bangkok Office. (n.d.). *Entrepreneurship education*. Retrieved from https://bangkok.unesco.org/theme/entrepreneurship-education

United Nations Conference on Trade and Development. (2022). *Entrepreneurship policy framework and implementation guidance*. Retrieved from https://unctad.org/topic/enterprise-development/entrepreneurship-policy-hub

University College London. (n.d.) *UCL innovation and enterprise strategy 2016–2021: Transforming knowledge and ideas into action*. https://www.ucl.ac.uk/enterprise/sites/enterprise/files/innovation-and-enterprise-strategy-2016-2021.pdf

University of California at Berkeley. (n.d.). *The 2018 strategic plan*. https://chancellor.berkeley.edu/sites/default/files/strategic_plan_overview.pdf

University of Queensland. (n.d.) *Entrepreneurship strategy 2018–2022*. https://ventures.uq.edu.au/files/3440/EntrepreneurshipStrategy.pdf

University of Waterloo. (n.d.). *University of waterloo strategic plan 2020–2025: Connecting imagination with impact*. https://uwaterloo.ca/strategic-plan/sites/ca.strategic-plan/files/uploads/files/c017209-stratplan2019-simple-oct316_accessible-ua_59906.pdf

World Economic Forum. (2009). *Educating the next wave of entrepreneurs: Unlocking entrepreneurial capabilities to meet the global challenges of the 21st century*. Retrieved from https://www.weforum.org/reports/educating-next-wave-entrepreneurs/

World Economic Forum. (2014). *Entrepreneurial ecosystems around the globe and early-stage company growth dynamics—The entrepreneur's perspective*. https://www3.weforum.org/docs/WEF_II_EntrepreneurialEcosystemsEarlyStageCompany_Report_2014.pdf

Xu, X. Z. (2019). Strategies for constructing ecosystem of innovation and entrepreneurship education in colleges and universities in the period of transformation and upgrading. *Research in Educational Development, 39*(Z1), 102–108. [In Chinese].

Yu, C. X., & Mei, W. H. (2021). Students' perception on the effectiveness of entrepreneurship education programs in China. *Entrepreneurship Education, 4*(3), 233–250.